# THE UNITED STATES
SPACE FORCE

# THE UNITED STATES SPACE FORCE

## Space, Grand Strategy, and U.S. National Security

LAMONT C. COLUCCI

Praeger Security International

An Imprint of ABC-CLIO, LLC
Santa Barbara, California • Denver, Colorado

Copyright © 2023 by Lamont C. Colucci

All rights reserved. No part of this publication may be reproduced, stored in a retrieval system, or transmitted, in any form or by any means, electronic, mechanical, photocopying, recording, or otherwise, except for the inclusion of brief quotations in a review, without prior permission in writing from the publisher.

**Library of Congress Cataloging-in-Publication Data**

Names: Colucci, Lamont, author.
Title: The United States Space Force : space, grand strategy, and U.S. national security / Lamont C. Colucci.
Description: Santa Barbara, California : Praeger, [2023] | Series: Praeger security international | Includes bibliographical references and index.
Identifiers: LCCN 2022039016 (print) | LCCN 2022039017 (ebook) | ISBN 9781440874833 (cloth) | ISBN 9781440874840 (ebook)
Subjects: LCSH: United States. Space Force. | National security—United States. | Outer space—Strategic aspects.
Classification: LCC UG1523 .C657 2023 (print) | LCC UG1523 (ebook) | DDC 358/.800973—dc23/eng/20220919
LC record available at https://lccn.loc.gov/2022039016
LC ebook record available at https://lccn.loc.gov/2022039017

ISBN: 978-1-4408-7483-3 (print)
      978-1-4408-7484-0 (ebook)

27 26 25 24 23   1 2 3 4 5

This book is also available as an eBook.

Praeger
An Imprint of ABC-CLIO, LLC

ABC-CLIO, LLC
147 Castilian Drive
Santa Barbara, California 93117
www.abc-clio.com

This book is printed on acid-free paper ∞

Manufactured in the United States of America

To my wife, Kathryn, and our children Isabella, Alfred, and Roland.

# Contents

*Preface* ix

*Introduction* xi

1. America at the Crossroads of Grand Strategy 1
2. History 25
3. Grand Strategy and Space Dominance 41
4. Space Intelligence 53
5. Allies, Diplomacy, and International Law 59
6. The Triplanetary Economy 67
7. Strategic Competition, Great Powers, Threats, and Enemies 75
8. The Service and the Mission 91
9. Pax Americana and Pax Astra 99

*Bibliography* 131

*Index* 149

# Preface

Like so many of Generation X, I grew up with the original *Star Trek* as an entry point. More important than the genre of science fiction was the topic of space. My generation then became the children of the original *Star Wars* trilogy of the 1970s and '80s. There is much merriment to be made pitting *Star Trek* fans against *Star Wars* fans. I have participated in this myself and am ambivalent at best toward the "hardcore" fans whose entire life centers around a work of fiction. I am even more negative toward the naysayers who view the entire genre poorly. To date, I have yet to meet many serious Space Force visionaries who are not "Trekkies" in some way, shape, or form.

However, this is not a book about science fiction, and it is not a book only about Space Force. It is about the synthesis among a new American military branch, our overall strategy, our vision of space, and a new economic revolution that will change the planet's destiny. It is also about those dark forces that seek to overtake America in this quest and create a dystopia of violence and chaos.

This preface is being written during the 2022 Russian invasion of Ukraine. Whatever the outcome of this conflict, it simply reinforces what I have been writing and speaking about for two decades. Great power conflict is not back. It never left. Russia has its own set of vital and national interests. These are often and will often be in direct competition with those of the United States. The same goes for China. The Russian invasion also illustrates another theme. This is the problem of the lack of strategic vision, in particular, grand strategy, in American policy circles. The myopic policy missteps of the Obama years, which were rife, is a bill that is being collected. It is not the purpose of this work to outline those problems. However, it should remind us that we need to think in terms of long-term strategy when it comes to space. Suppose we use the same tired models in American foreign policy and military strategy. In that case, we will lose the initiative in space, and if we do that, we may never be able to recover. Any strategy that does not place American primacy at its heart is a failed strategy.

I want to give a special thanks to those space and national security professionals who agreed to be interviewed for this book: General John Raymond, chief of space operations, U.S. Space Force, and member of the Joint Chiefs; Dr. Joel Mozer, chief scientist of the U.S. Space Force; Lt. General John Shaw, deputy commander of U.S. Space Command; General (RET) Robert Kehler, former commander, U.S. Strategic Command; Lt. General (RET), Steve Kwast, former commander of Air Education and Training Command and president of Genesis Systems; Brig. General (RET) Simon Worden, former director of development and transformation, Space and Missile Systems Center; ambassador and former director of Central Intelligence Jim Woolsey; ambassador and former director of the Strategic Defense Initiative, Hank Cooper; former Speaker of the House Newt Gingrich; Lt. Colonel (RET) and Dr. Coyote Smith, professor of strategic studies; Lt. Colonel (RET) Peter Garretson, former chief of the air force future technology branch; Rick Tumlinson, the "godfather of the space revolution" and author of the term NewSpace; Josh Carlson, space strategy consultant; Dr. Namrata Goswami, space policy consultant; and Ian Molony, space architecture, engineering, and spacecraft design.

# Introduction

In 2008, I was touring Cape Canaveral and disembarked near the pad where the Apollo 1 accident occurred. Three American heroes died in a horrible fire on January 27, 1967. As I neared the location of the disaster, I noticed what appeared to be a wreath and American flag lying in the dirt, like it was debris. I was astonished to see our guide and the rest of the visitors walk by in oblivion. In anger, I stalked toward the wreath and flag and propped it upright onto its thin metal frame. My immediate thought was, "Is this what our space program has become?" However distant it becomes with each passing year, we must keep alive a certain memory. First is that of the tragedy of Apollo 1. Second—that of the Apollo 11 moon landing. And with that, the famous transmission, "Houston, Tranquility Base here, the Eagle has landed." This was done by Americans, with American ingenuity and, more importantly, with the kind of foresight we need today.

As you are reminded of that day, July 24, 1969, remember that the space program and manned space exploration specifically are one of the many keys to America's future, not only as a global superpower but also as the leading economy. The two cannot be separated, and neither of them will have a future without America leading the way. In the here and now, not in some murky future. It is precisely because of the numerous twists and turns in our economy, the threats posed by other great powers and rogue states, that this is the time for such a clarion call. This time needs to be capitalized on to advance the absolute need for a renewed American commitment to space. The country that makes this commitment will have a secure future.

How do we do this? First, the American government must prioritize this to the electorate and make an ideological and practical case for space and the new Space Force. The first time I mentioned this to a group of space professionals, I received quizzical and perhaps skeptical looks. Why was I suddenly talking about the American voter in a talk about Russian and Chinese national security strategy? This made me think that I had hit a nerve. Space professionals obsess about all things space. This is not a surprise. However, few understand American politics, and many focus on space to insulate them from partisanship and the political scrum. I think many even disdain political talk, probably out of a spirit of fear. If America is serious about space, it better make sure the electorate is as well. Our political system dictates that any prolonged policy without public support will fail. This is also not hard. There is nothing inherently partisan about space in the vast middle of American politics. The fringes won't like space, and they won't like Space Force because of various extremist opinions that ultimately die of their own weight. However, if the space professional community ignores the electorate or, worse, takes them and their tax dollars for granted, we should close up shop now. In other words, this connection to the American electorate needs to be a central priority, not one on the periphery. It should not be left to public relations "experts." It needs to be part and parcel of the everyday aspect of the Space Force's job. One of the core rules of politics, if you don't define yourself, others will do it for you.

On the romantic side, we need to hearken back to President Kennedy demanding that America and Americans must lead this human endeavor, that the banner of freedom and democracy must be at the forefront, and that it is not only our challenge but our duty and responsibility. If not us, then who? If not now, when? The American government needs to make the national security and economic case in stark and clear terms on the practical side. The cost of both, for another power to supersede us, would be catastrophic at every level.

Readers may also benefit from knowing that I have been fortunate enough to be involved directly with Space Command and Space Force. I am an unabashed proponent of them and view their success as a success for the United States. My involvement has been giving briefings and presentations to them and being one of the authors and contributors to both Space Futures 2060 and Space Futures 2045.

I am not a futurist and will leave that field to those experts, many of whom are critical to forward-thinking on American strategy and the U.S. Space Force (USSF). This book is a synthesis combining the role of the new USSF, American national and grand strategy, and the new economic revolution. There are many experts in each of these realms, but few, if any, have sought to combine them into a whole. The Space Force is busily defining itself and fighting off the initial challenges of an embryonic institution. As has unfortunately been the case, American national security is consumed by fighting the fires of today and perhaps the near term with scant attention to how decisions now will create the conditions of grand strategy in the future. Finally, the new economic revolution is dominated by "new space" advocates, primarily from the private sector, who are often ambivalent, ignorant, or in some extreme cases, uncaring about the military, security issues, and great power conflict.

This book addresses this and tries to offer insight into why we are here and where we are going—questions that guarantee future American primacy.

# CHAPTER 1

# America at the Crossroads of Grand Strategy

The 1896 novel *Quo Vadis* highlighted the question "Where are you going?" The meaning of this question goes far beyond the text and forces the person to confront a host of other questions, wrestling with a simple geographical answer of the most profound trajectory of the mind and spirit. America is at such a "Quo Vadis" crossroads, and its choice will determine whether it remains the premier power, sinks into being a great power among many, or worse. Anything less than the first choice dooms the United States, the American people, and Western civilization to conflict, violence, and possible subjugation. American national security will be determined by its superiority in the space domain through the actions of the U.S. Space Force. This book illustrates that space policy is subordinate to national security policy, which is a servant to American grand strategy. We have compartmentalized ourselves into disunity and handed over our destiny to a technocratic and bureaucratic corps that has lacked vision, historical understanding, and future thinking.

American society is a reflection of this. How often does a person get shifted from one person to another, each claiming that they are not the responsible figure? How often are the ultimately accountable figures unable or unwilling to make the tough decisions? How do they usually inform us that there are multiple turfs and lines of control that they can't cross? How often does something from the most mundane to the most important get derailed or delayed? This last one could mean that something won't get done in a timely manner, a loss of efficiency. In the most extreme cases, people die, sometimes many people. Every second that America failed to develop a coherent policy regarding Syria, hundreds and then thousands of people died. This is the cost of bureaucracy and technocracy. This no longer represents an America ruled by liberty under law, but a society of the law of rules. This model must be placed in the ash can of history where it has always belonged if we are to win the space race. The streamlining and destruction of silos that will be necessary is so titanic that it will cause significant disruption, and many political fiefdoms will be threatened. The stovepiping is so extreme in D.C. that it threatens national security. I will be the first to acknowledge and support the Founders' belief that overly centralized power is a threat to republican values and enhances the chance of the tyranny of the few or the one. However, let us also remember that the other fear that the Founders had was the tyranny of the many and their allies in a permanent political, bureaucratic class.

America has "always been a spacefaring nation" (Raymond 2021). If we wish for the Pax Americana (the system of order created by the United States) to continue, it must embrace the Pax Astra (a new system of order that will govern space and earth).

If national security strategy is dependent on grand strategy, what can we do in space? First, we must have a national space strategy synthesizing multiple ideas and fields. It must have an intellectual foundation that rests on accomplishing an American destiny in space. Second, it will need to wrestle with the numerous political challenges that are not only about political party, personality, and ideology but also a rising progressive movement that does not wish for American primacy in the future. Third, rules for space, space governance, and activity must be molded along with the "international liberal order" that has governed American-led international norms since the Second World War. This will require a firm commitment by Western powers to stand with the United States as it leads humankind into the twenty-first century. Equally, American diplomatic efforts must shore up the various alliance structures between NATO, ANZUS, Japan, South Korea, Israel, Taiwan, and other nations that wish to join the American-led reach for the stars. This destiny will only be possible by developing the space organizational structures and institutions to achieve it. It will primarily be accomplished by the new United States Space Force (USSF). Finally, America must lead a new economic revolution that embraces space commerce, resource harvesting, colonization, exploration, transportation, surveillance, logistics, and research. None of this can be accomplished with a single administration's space policy. Still, the embrace of a space strategy that takes the theories of space power to create space policy under the guidance of American primacy is necessary.

Diplomats often talk about *interests*: vital, national, and peripheral. Much like the misuse of terminology over strategy, there is misuse over the terms *vital interests* and *national interests*. No discussion of national security is possible without an understanding of both. Vital interests are existential interests whose failure to protect could bring about the extinction of that civilization. These are the interests on which a state is unwilling to compromise. In essence, vital interests measure the survival of the state, and its core culture and institutions, as well as its morality and honor. Although there will always be some debate about vital interests, they tend to be stark, raw, and blatant; they stare a nation in the face with a gaping maw ready to devour those that ignore them. The costs of vital interests are severe and catastrophic. Vital interests demand the highest priority of policymakers, with the price of annihilation if they are flouted. The nuclear threat posed by the Soviet Union to the United States is a perfect example of this in modern times.

*Vital interests* are existential. They mean survival. For example, the United States had a vital interest in what the Soviets did with their nuclear weapons.

*National interests* are critical for a nation's military position, economic prosperity, and stability. National interests cause more debate; there is less consensus because the stakes are lower, albeit incredibly important. Some have created a more complicated tier of interests among vital, critical, and serious: this only further illustrates the need to create such priorities in the first place. However, national interests consume much of national security strategy. A nation must create its forward-looking goals and objectives based on this, and on this alone. In an attempt to create a clinical definition of national interests, Bruce Jentleson cited the four Ps of national interest: power (military, alliances); peace (diplomacy and international institutions); prosperity (trade); and principles (democracy, democratic peace) (Jentleson 2010). National interests are those that a nation defines to achieve national objectives. These objectives concern power, economy, and morality, depending on the nation. Clearly, in the American context, all three have played titanic roles in determining national objectives, but unlike most nations, American national interests have always tried to combine them. The successful national security doctrines are those that do exactly that—they unite the geopolitical, the commercial, and the just to create a whole. For example, Japan has a national interest in the free flow of oil through the Persian Gulf.

*Peripheral interests* are those that you like or want, but are not critical. For example, the United Kingdom has a peripheral interest in human rights in its former colonies. This is in no way discounting them. It is merely a clinical way to divide those that are existential (vital), needed (national), and wanted (peripheral).

It will be apparent that a total commitment to space expertise and personnel will be required to achieve this. Professor of Strategic Space Studies Coyote Smith stated in his defense of a Space Corps, "The expertise required to operate in space is so fundamentally different and specialized, and the space-minded perspective so altogether different, that it will never serve the nation to have this expertise and perspective under-resourced and buried under another service" (Smith 2017). Yet, if we fast forward only three years from this article to 2020, we see the Department of Defense endorsing the vision of many space power advocates.

The Department is taking innovative and bold actions to ensure space superiority and to secure the Nation's vital interests in space now and in the future. Establishing the U.S. Space Force (USSF) as the newest branch of our Armed Forces and the U.S. Space Command (USSPACECOM) as a unified combatant command, as well as undertaking significant space acquisition reform across the DoD, has set a strategic path to expand spacepower for the Nation. (Department of Defense 2020a)

This document begins to flesh out the vision that American space power advocates have been evangelizing for decades. It recognizes the fact that it has gone missing for years by many space enthusiasts. Space is in the vital national interest of the United States and is the only way to ensure such national interest broadly. This DOD report maps out three goals: maintain space superiority; provide space support to national, joint, and combined operations; and ensure space stability (Department of Defense 2020a).

One of the other critical items that this report highlight is the declaration that space is a "distinct warfighting domain." In other words, we are abandoning the fashionable fallacy that space is or could remain a sanctuary from earthbound rivalries. Instead, if one wishes to embrace peace and order, it must be done under American auspices.

## SPACE AND AMERICAN NATIONAL SECURITY

Space policy is the ultimate high ground and will be synonymous with national security. It "underpins every instrument of national power" (Raymond 2021). It will be the wagon that provides new military and economic power. It will also be the more complicated measure of national prestige. It is a place for America to "put forth grand ideas" (Garretson 2021). Space power will simply be the highest benchmark of military power over everything else (Gingrich 2021). It provides our "maximum field of view" (Moloney 2021). Currently, space is critical. Our space capabilities grant us over-the-horizon communication to the battlefield (Mozer 2021). Today and in the future, space technology offers the ultimate military offset (Cooper 2021) that assists America in blunting quantitative threats.

Readers should be reminded about the "offset" strategy. American national security has recognized three eras of the offset strategy. An offset strategy overcomes a nation's deficiency by countering it with an asymmetrical approach. If you face an opponent with an overwhelming number of fixed fortifications, like the Maginot Line in 1940 France, Germany did not try to go through it with infantry lines. Instead, they "offset" the French advantage by using amassed concentration of tanks, many of which went around the fortifications. Following the Second World War, the United States realized that it would not counter enemies with equal or superior numbers. That was a losing strategy. So it decided to find offsets. The first offset was under the Eisenhower administration. The administration knew it could not counter Soviet and Chinese numbers, so it developed the "New Look" and "Massive Retaliation." Nuclear weapons would be the offset the American inferiority of numbers. The second offset of the 1970s and 1980s focused on precision weapons, stealth, firepower, and long-range targeting. The third offset, beginning around 2014, ushered in emphasis on artificial intelligence, miniaturization robotics, cyber, and data. Some are calling for a fourth offset centered around alliances and partnerships.

It may be a mistake to view space as the ultimate offset. Although space offers the ultimate in the American quest for more superior technology and potential firepower, it requires a complete rewiring of national security strategy. Space power is not so much an offset as it is the strategy writ large.

It enables "the American way of warfare" (Kehler 2021). It is currently our most advanced military asset and our most vulnerable and unprotected arena. It is, in the words of Coyote Smith, our Achilles heel. "It is America's largest center of gravity because every sector supporting the instruments of national power—diplomacy, information, military, and economic—has grown dependent upon space forces for daily operations, with little or no alternative means. Space has become America's Achilles' heel because those satellites are undefended" (C. Smith 2020). Most Americans don't realize how far-reaching our space dependencies are "from everyday life to warfighting" (Shaw 2021). It is the driver for the internet, communications, weather forecasting, surveillance, reconnaissance, missile warning, and overall intelligence. The United States cannot simply monitor the world without our space assets (Woolsey 2021). In the future, space power moves us from the three-dimensional realm to the fourth that will take us from the orbits of the Earth to Mars (Kwast 2021).

## SPACE FUTURES WORKSHOP

In September 2019, the then-U.S. Air Force Space Command (before the re-creation of U.S. Space Command and creation of the Space Force) sponsored a workshop titled "The Future of Space 2060 and Implications for

U.S. Strategy." As one of the many authors, presenters, and participants, I can attest to this exercise's value for the future of American national security.

We titled the best outcome for the United States (out of eight future paths, two of which would spell catastrophe for the United States) "Star Trek." The report argued for three needs: the need to maintain the United States and Allied space power, protect and defend U.S. interests in space, and keep the space "commons" free. Three questions were addressed: Can we make money in space? Can we live in space? And will the United States and its allies control space? This 2060 "Star Trek" future was one where there would be a robust human presence in space. There will be an enormous economic opportunity, and, most importantly, it will be led by the United States and our alliance partners.

This endeavor combines futurist thinking, technology trends, and geopolitical realities. The futurist concept requires one to suspend current thinking and expand into areas not firmly planted. It does not attempt to predict the exact future but opens the mind to future possibilities.

This thinking is based on the Futures cone. As one can see in Figure 1.1, this includes the probable, the preferable, the plausible, and the possible.

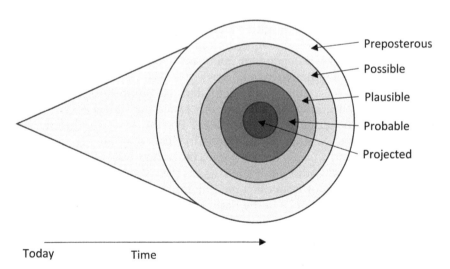

**Figure 1.1** Futures Cone

*Source:* Created by author, based on ideas from https://thevoroscope.com/2017/02/24/the-futures-cone-use-and-history/

This inaugural workshop focused on the challenges from peer rival China. Though focused on a single rival, these alternate futures align with those developed in the Air Force Strategic Environmental Assessment (AFSEA) 2016–2046. From this scenario-based process, the workshop developed and examined eight future scenarios across the following axes:

- **Human Presence:** *How broadly and in what numbers do humans live and work in space?*
- **Commercial Potential:** *What is the economic level of new, persistent revenue from space activities?*
- **Leadership:** *To what degree does the U.S. coalition lead in the creation of the civil, commercial, and military space capabilities and in establishing the norms, standards, and international regulations in space? (e.g., based on free world concepts of a liberal international order?)*

Eight scenarios, or alternative futures to 2060, were derived by defining realistic bounds as to the minimum and maximum developments along three axes of space power: human presence in space, economic importance of space, and U.S. coalition leadership in space. For this analysis we assumed that the states along each axis are independent of those along the other axes. The upper and lower bounds of each axis are summarized below. The bounds are derived from the SFW brainstorming activities informed by the assumptions as to the global state and the extrapolation of current space trends to 2060. The Annex contains detailed information of these trends and assumptions, especially relating to China's space program. The methodology used to determine future scenarios is also presented in the Annex.

**Human Presence:** *How broadly and in what numbers do humans live and work in space?*

### Upper 2060 Bound

- Thousands of individuals from many nations live and work from low Earth orbit (LEO), across the cislunar, Moon, and Mars regions.
- Increased human presence supports civil, commercial, and military needs and interests.
- There are one or more crewed bases on the Moon supporting science or economy.
- There are crewed habitats across cislunar space supporting the overall logistics for space manufacturing, tourism, and resource extraction.

- Habitats in space, the Moon, and Mars are progressively constructed and maintained using extraterrestrial resources.

**Commercial Potential:** *What is the economic level of new, persistent revenue from space activities?*

**Lower 2060 Bound**

- The space economy advances marginally beyond short-term projections of the current state.
- The economic value of the space economy is some small trillions of dollars and represents at most 1–2 percent of the global economy.
- The space economy is confined primarily to LEO/geosynchronous Earth orbit (GEO) and principally supports terrestrial needs for communications; positioning, navigation, and timing (PNT) capabilities; information gathering; and a low level of tourism.

**Leadership:** *To what degree does the U.S. coalition lead in the creation of the civil, commercial, and military space capabilities and in establishing the norms, standards, and international regulations in space? (e.g., based on free-world concepts of a liberal international order?)*

**Lower 2060 Bound**

- Many national, international, and transnational commercial interests operate in space.
- Leadership in space is not based on an extension of the fundamental terrestrial tenants of a liberal international order.
- The U.S. coalition is not the space leader in that they are at a serious disadvantage in protecting their interests and capabilities in the civil, commercial, and military realms.

**Upper 2060 Bound**

- Overall, the space economy is expanding rapidly, contributing at least 10 percent of global gross domestic product (GDP) with wide, diverse participation from nations.
- The economy includes power, planetary communications, global information services, manufacturing, resource extraction, and tourism.
- Major industrial capacity for power, resource extraction, and manufacturing has developed, driven primarily by terrestrial demand, but increasingly to support the space economy.

- Space tourism is a major industry available to a wide cross section of the public.

**Upper 2060 Bound**

- The United States alone or with its allies holds the lead power position in aggregate across the civil, economic, and military realms.
- Space is protected as a free domain under a rules-based, international order with established norms of behavior that promote the philosophy of open trade and space as a commons for all of humanity For this analysis we assumed that the states along each axis are independent of those along the other axes. These combinations of bounds produced eight future scenarios, shown graphically in Figure 1. Maximized values of the axis bounds are shown in dark gray symbols. Minimized values are shown in open symbols. Scenario titles are derived from the predominant characteristics of the scenario. Green icons are those generally favorable to the United States, while orange and red represent those generally unfavorable to the United States.

These three scenarios assume a major increase in the importance of space globally, with the U.S. coalition retaining space leadership across the civil, commercial, and military realms. Further, they broadly share similar characteristics in these realms. Brief descriptions of these scenarios, as well as key resulting characteristics, are provided below.

**Star Trek:** *Most optimistic and expansive*

The U.S. coalition retains leadership over the space domain and has introduced free-world laws and processes that have led to significant global civil, commercial, and military expansion in space and resulted in large revenue streams. Thousands of humans live or work in space at a variety of habitats across cislunar space, the Moon, and Mars.

**Garden Earth:** *Optimistic and expansive*

The U.S. coalition retains leadership over the space domain and has introduced free-world laws and processes that have led to significant global civil, commercial, and military expansion in space and resulted in large revenue streams. However, human presence is limited and most processes are controlled remotely or robotically.

**Elysium:** *Optimistic and expansive*

The U.S. coalition retains leadership over the space domain and has introduced free-world laws and processes that have led to significant global civil, commercial, and military expansion in space. Thousands of humans live or work in space at a variety of habitats across cislunar space, the Moon, and Mars. However, large revenue streams have yet to materialize. Commercial activity is focused in LEO to GEO terrestrial communications, information, PNT, and to provide key parts of the civil and commercial infrastructure required for the continued expansion of human presence in space. These three future scenarios posit a major growth in the importance of space and share the characteristic of an alternate to a U.S. coalition as the leader across the civil, commercial, and military elements of space power. They further posit a significant power advantage of this lead space power over the U.S. and its allies. While who might develop to be this leading space power is uncertain, we have chosen names with a Chinese reference since China is presently the most likely candidate. In these futures, the alternate lead space power views the U.S. coalition as a rival at best and a potential adversary at worst.

Negative Futures: Zheng He, Wild Frontier, and Xi's Dream

**Zheng He:** *Expansive but most pessimistic*

An alternate nation exercises leadership over the space domain and has introduced laws and processes that promote their interests or limit the actions of rivals. Leveraging their growing technological edge and using noncompetitive practices, they attract a growing, disproportionate share of global space revenue streams. Thousands of humans live in space to maintain lunar and Mars bases to promote national prestige, further patterns of dependency, and support the technology and infrastructure for commercial and military space leadership.

**Wild Frontier:** *Expansive but pessimistic*

No clear space power exercises leadership over the space domain. However, the growth in space capabilities of national and private entities has resulted in global civil, commercial, and military expansion in space and led to large revenue streams. However, human presence is limited, driven primarily by national prestige, exploration, and tourism.

**Xi's Dream:** *Expansive but pessimistic*

An alternate nation is the lead space power, though the importance of space is driven by the increased human presence in space for exploration, tourism, and to support and maintain commercial space capabilities. Large revenue streams have yet to materialize. Commercial activity is focused in LEO to GEO terrestrial communications, information, PNT, and to provide key parts of the civil and commercial infrastructure required for the continued expansion of human presence in space.

## KEY RESULTING MILITARY, CIVIL, AND COMMERCIAL CHARACTERISTICS

### Military

- The alternate chief space nation and its allies lead in military space power and have the range of military capabilities necessary to:
    - Defend the critical elements of their civil, commercial, military, and human space assets, as well as C4ISR (command, control, communications, computers [C4] intelligence, surveillance and reconnaissance) to monitor and control space operations and provide information services in, through, and from the cislunar environment during peace and conflict.
    - Project power to achieve space superiority across the cislunar domain and across the full range of conflicts in space or that extend to space.
    - Leverage their commercial and civil leadership to maintain and further their lead in military space power. They use their commercial lead to exclude other spacefaring nations from critical locations and resources in space. They use their cost advantage from their space commercial infrastructure to outproduce other nations' space military capabilities.
    - Further their civil and commercial leadership by exerting pressure or by threats of force.
    - Restrict their unilateral military actions to avoid the development of alliances among the nations holding the preponderance of military space power. The U.S. coalition works to create such military alliances.
- The United States and other spacefaring nations are forced to create independent infrastructure to ensure national interests during peace and conflict.
- Alternatively, the dominant player(s) do not hold the preponderance of military space power. This restricts their ability to act

unilaterally in space. The United States and its allies work to create alliances to counter their lead.

### Civil

- The alternate chief space power promotes norms of behavior, rules, and laws for space that serve their self-interests. Where that's not possible, they exploit the diversity of interests of spacefaring nations/entities to limit the establishment of space norms, rules, and laws that impact their lead position or limit their range of actions.
- They lead in developing multinational space, civil infrastructure to promote alliances, to exert infrastructure control to their benefit in peace and conflict, and to establish patterns of dependency for other terrestrial and space powers/entities.
- The preponderance of civil space power is distributed among a large and varied group of spacefaring nations/entities, limiting the ability of the alternate chief space power to act exclusively in its self-interest.
- The U.S. coalition forms alliances to oppose the lead power's actions and have built alternative, civil space infrastructure to support their nation's interests.
- Multiple nations have established bases/colonies on the Moon and Mars and pursued asteroid exploration, but the alternate chief space power leads in civil exploration to promote national prestige, further patterns of dependency, and support the technology and infrastructure for commercial and military space leadership.

### Commercial

- While the preponderance of commercial power is distributed across a number of nations with the U.S. coalition having a significant space commercial presence, the alternate chief space power has a large and growing technological and market-share advantage over all other spacefaring nations that attracts a disproportionate share of global commercial space investments.
    o They have a significant advantage in scale and pricing for space structures (satellites, habitats, etc.), power, and low-cost and flexible launch.
    o They lead in space resource extraction to meet critical space and terrestrial needs for raw materials, providing them commercial and political leverage.

o They have de facto control over key locations in space (Lagrangian points, etc.) and on the Moon (polar region, space elevator lunar point, key mineral deposits, etc.).
   o They have a commercial space enterprise increasingly independent of systems and elements produced on Earth.
- They use noncompetitive practices to maintain and extend their lead position, with limited capability by the U.S. coalition or other nations to counter these practices. They limit these actions as required to avoid encouraging alliances to oppose them.
- The United States subsidizes U.S. space industries supporting critical national needs where supply by the lead space power or its allies pose unacceptable risks (U.S. Space Command 2019).

As one of the presenters and authors of Space Command's "The Future of Space 2060," we feel it is incumbent on those enmeshed in space strategy to comment during this genesis period for the new U.S. Space Force.

Space Futures is now in Phase II. As one of the lead team members, we are trying to prepare for the "Space Force after next" around the year 2040 by focusing on possible scenarios entitled "Images of Future Operations (IOFOs)."

The six scenarios developed for the workshop are colorfully titled:

1. *A Slip in Perseverance*—Loss of U.S. leadership in an expanding and increasingly valuable space domain. What happens when space is rich and hospitable, the United States is not the one running the place, and other space powers are achieving infrastructural breakthroughs?
2. *Claiming Paradise*—Sudden transformation in space value driven by black swan technological breakthrough. Warp drives. Extremely rapid propulsion and traditional propulsion and launch technologies coexist in a world where the space economy did not live up to its promises . . . yet.
3. *A Crowded Garden*—Conflicts in an expanding and highly multipolar space world. Multiple great powers within a flourishing space economy compete in the unforgiving terrain of gravity.
4. *Venus and Mars*—Piracy, revolution, and conflict in a unipolar space world. A blossoming space economy reveals the limits of defending an unconstrained spatial domain.
5. *The Expanse*—Cascading disasters strain actions and space power relations across an extended and expansive space domain.
6. *A Great Leap Backwards*—Failed international order in space in a deadly series of man- and nature-triggered space catastrophes imposes colossal pressure on the space enterprise across the globe, uncovering buried tensions.

**TERMS**

Space Force is under siege by various factions with drastically different ideologies, ranging from left-wing secular progressives who believe Space Force will militarize an already militarized space to libertarians who believe that it is a colossal waste of money. These two vastly different groups share two traits: first, they want to characterize Space Force as a "Trump vanity project." USSF has become the victim of partisan politics, equating disliking President Trump with disliking Space Force, regardless of its merit (Broad 2021). Second, they have zero understanding of the strategic precipice America is walking on. These groups focus on cosmetic attacks because they are entirely out of their depth on strategy. They mock the uniforms, the name "guardians," the use of the delta insignia (which predates *Star Trek*), and the general absurdity via a Netflix series.

It would be an excellent exercise to review basic terminology for the reader as space professionals often use these terms without understanding that even the educated public has little knowledge of these ideas. **Grand strategy**, at its core, attempts to harness military, economic, and political power to advance the nation. It is created organically, over decades and centuries, and for it to be successful must be forward-looking, peering across the horizon into the centuries uncounted. Grand strategy is the most critical form of statecraft. It implies the use of force to promote these interests. Grand strategy is married to hard power and military force; unlike domestic policy, it creates the conditions for either total triumph or total destruction (Colucci 2018). There is confusion over grand strategy and foreign policy. However, grand strategy is primarily a vehicle of military power over a long arc of time. Grand strategy is not only based on military power, but the absence of military power makes any attempt at such strategy futile. This military dimension illustrates again the need for grand strategy created by national security doctrines, which hold the reins of leadership instead of the reverse logic of allowing tactical considerations to dictate terms. The contemporary U.S. military often seems hobbled by being charged with winning battles by losing wars, and with winning wars but failing to attain national security objectives by disregarding the lessons of history and thus duplicating past mistakes. Grand strategy must also be based on reality and not purely on unrealistic ambition. It is often so deeply rooted in the history and substructure of the nation that policymakers and generals are not even aware they are serving it. It should also be remembered that national character affects a nation's strategic decisions. If grand strategy is to be effective, it must be forward-looking and long-term. **Geopolitics** is the study of international affairs and geography, and its sibling is **astropolitics**, which replaces geography with space. "Astropolitik is grand strategy. Indeed, it is the grandest strategy of them all" (Dolman 2002, 1). Both of these are dominated by security studies and national security implications. A **great power** is any nation (not state) that exercises considerable influence over geopolitics and astropolitics. The United States, as the dominant geopolitical player, is the order maker, the hyperpower, one step above a superpower. This exploration aims to ensure that the United States remains the order maker for astropolitics as well.

As far as space terminology goes, one should limit the technicalities to a bare minimum. One of the debates is when does "space begin." There is disagreement here, with some arguing that earthbound objects exist from the surface to the atmosphere. Many endorse the Karman line, which is about 62 miles above sea level (Drake 2018). This debate may become more academic as space becomes more contested. Objects in **low Earth orbit** (LEO) travel between 99 and 1,200 miles above the surface of the Earth and have an orbital period (the time it takes for the object to orbit the Earth of between 88 and 127 minutes). LEO is where the majority of manmade space technology currently exists, such as the International Space Station. **Geosynchronous equatorial orbit** (GEO) is where objects are in high orbit above 22,000 miles match Earth's rotation (24 hours), which is helpful for communications and surveillance satellites. **Cislunar** space is the space between the Earth's atmosphere and the area right beyond the orbit of the Moon. Strategically, cislunar includes the Lagrange points, which are the points in space where there is an equilibrium between Earth's and Luna's gravitational force. Further out, there is **solar system**, which is space beyond the Moon's orbit. Finally, there is the term **NewSpace**, coined by Rick Tumlinson in 1998, is "any company that is created by, funded by, markets to, profits from or supports the opening of the High Frontier to humanity" (Tumlinson 2020). These companies will be the primary fuel of the space economy. NewSpace visionaries like Rick Tumlinson saw the possibilities for space early on. Many of them have been critics of the NASA way of doing things. In other words, many of the original NewSpace advocates early on saw the economic and human presence (colonies as an example) potential long before the twenty-first century. They were already talking about asteroid mining and fuel extraction when many others mocked such ideas.

## BASIC ISSUES

Although, as General Pete Worden explains, "Space represents an unlimited economic and physical future for mankind" (Worden 2021), there were and are many varied positions. Further, a review of the fundamental issues is in order. First, there were long-standing debates about the need for a military space branch. Second, these debates distilled to arguments over the technicalities and concepts of "Guard," "Corps," and "Force." The third idea, a Space Force, would be an independent branch after transitioning from the U.S. Air Force. These proposals would have pushed the United States down the right path, at least partially. One idea was that of a Space Guard, modeled on the Coast Guard, who would oversee civil and commercial space activity and ultimately deal with problems ranging from search-and-rescue to planetary defense. This is a softer, somewhat subordinate role than other proposals. In the middle was the idea of a Space Corps, modeled on the Marine Corps; the Marines are technically subordinate to the Navy, but it is an autonomous service. A Space Corps would likely be under the jurisdiction of the U.S. Air Force. However, although both of these plans were much better than the pre-Space Force environment, they are far from integrating space strategy into national security strategy and grand strategy. This could only be achieved through a separate military branch, which should be titled the Space Force. Ultimately, the Space Force (USSF) and Space Force personnel took the moniker of Guardians.

## BASIC PROBLEMS

In many ways, the United States is a victim of its own success. We are the dominant power and the dominant space power, but potential adversaries appear to be catching up and perhaps surpassing us (Ziarnick 2015, 2). In essence, the fundamental problem for the United States is whether it wants to take an active decision to be the dominant space power or a reactive power, assuming that it will continue as the dominant power based on inertia. The United States may not have a chance to react to a Sputnik surprise if it is more like a Pearl Harbor. Many problems with American space strategy and policy have been highlighted over the decades following the Second World War. In 2017, Congressman Mike Rogers hosted a space symposium where he laid out four of the most significant problems. The first was that space organizations and decision-making was fragmented. Second, space is not given the priority it should. Third, America was not developing space personnel, and finally, the entire space program must be integrated (Rogers 2017). These were the fundamental problems that Space Force was supposed to address.

These debates were held among space and military professionals in niche areas and did not include the broader national security elite or the electorate. This is why the extremists have controlled the narrative and may offer cover to those in the new administration who want to stall Space Force rather than eliminate an already existing institution.

In an interview with the *New York Times* on March 8, General Raymond, U.S. Space Force Chief of Space Operations, summarized the new branch's importance: "I think it's really important for the average American to understand access to space and freedom to maneuver in space is a vital interest" (Swisher 2021).

The benefit to the American people protected by our Space Force Guardians is immeasurable. Any professional looking at the future of American national security realizes that this security will depend on which power exercises military primacy and space governance. Eventually, there will no longer be a separation between what we call national security and space strategy, to be precise. They will literally and figuratively be the same. The benefits and threats from space will dwarf those on Earth. Ultimately, space strategy will dictate all of those benefits and threats on Earth. The famous James Dolittle of the Tokyo Raid stated in 1959, "We, the United States of America, can be first. If we do not expend the thought, the effort, and the money required, then another, more progressive nation will. It will dominate space, and it will dominate the world" (Dolman 2002, 86). Dolittle, like many pioneers, such as those who led the Air Force in the early years of the service, were often mocked or sidelined by status quo forces within the military and the government as had been done to Billy Mitchell. The time is fast approaching when the mightiest carrier task force, tank platoon, or bomb squadron will be utterly vulnerable to space-based threats just as the medieval fortress became almost useless against mobile artillery and, later on, warplanes. This concept that both space power and national security are one in that they merge is monumental. It will be difficult for many to accept. However, it is not without precedent. Military power has long stretches of stability in broad strokes, followed by inertia, and then transition. Those powers that can't step from inertia to transition become subordinate to those who can.

In the beginning, where masses of warriors were thrown at the enemy, the Romans broke this model the best by organizing infantry in a way that not even the Egyptians, Greeks, and Persians could manage. The medieval period was dominated by armored cavalry and fixed fortifications. The combination of the long bow and then gunpowder shattered these assumptions. Sea power and breech-loading rifles dominated the period until the twentieth century when the machine gun, tank, battleship, and aircraft carrier appeared. We are now in a period dominated by the nuclear-powered aircraft carrier, advanced precision missiles, cyber, and special forces. However, the pivot point to space is now. The military that dominates space will dominate all military fields, making space power and national security the same.

Thus, a news flash for opponents or those that mock Space Force is this: space was militarized long ago and continues to be militarized. Russia and China fully intend to amplify this regardless of American efforts, including those of diplomacy and negotiations.

## BASIC ECONOMICS

The coming economic revolution, the revolution of "NewSpace," also called the "triplanetary economy," will unleash economic forces and powers that will extend economies and resources beyond anything in human history. This econosphere will either be protected by Space Force or left to be exploited by America's adversaries. The NewSpace economy will need to be protected, communication will need to be managed, travel and spacecraft control maintained, and debris will need to be cleared. No economic system can exist without preserving the law, private property, contracts, and protection from hostility, violence, chaos, and criminality.

The future of WMD defense, cyber defense, energy production, environmental protection, and democratic values will entirely depend on American space strategy.

Americans have dreamed of going to the stars for generations. The Apollo missions were thought to be the starting point for the United States to be a spacefaring people. Still, this dream drifted backstage as the political class allowed itself to be captured by the winds of popular culture and perceived expediency. In June 2018, President Trump resurrected this dream when he called on the military to create a new service branch called the Space Force. Although this gave new life into the dream, it also reignited the debates about Americans and space, especially the purpose of a new military branch.

While there has been ample discussion of the political, bureaucratic, budgetary, logistical, and technical challenges this poses, few have focused on how such an organization would fit into national security strategy, especially American grand strategy. Grand strategy is often ignored because it is inconvenient, hard to change, and subject to the tyranny of the status quo. Its development requires a formidable depth of knowledge. No electoral constituency holds a president accountable for not having a grand strategy even though having one is the *raison d'être* of the presidency. To ignore grand strategy is to engage in ad hoc policy anchored by nothing and moving nowhere. Grand strategy is further burdensome since it requires constant adaptation. American grand strategy is fundamentally based on military primacy, and space dominance will determine which nation is in that position.

## THE NAVY OF SPACE

One of the commonalities among Space Force advocates who view it as fundamental to the future of American national security is that their primary references are naval rather than aerial. This might strike many as odd since the vast majority of those in uniform wear the blue of the Air Force. However, this hearkening to naval strategy ranges from the cosmetic to the depths of strategic thinking. Cosmetically, many argue for the adoption of naval ranks, echoing William Shatner's (Captain Kirk of the *Enterprise* in classic *Star Trek*) famous article entitled "What the Heck Is Wrong with You, Space Force?" He argued that Space Force should adopt naval ranks to align with the tradition of exploration and history (Shatner 2020). Here lies another place where there is a synthesis between space and our civilization and culture and the science fiction genre.

However, this attention to naval analogies goes much deeper than cosmetics or references to Star Fleet. It often revolves around the thinking of

Alfred Thayer Mahan, who is the center of gravity among strategic "navalists." Those in the nineteenth and twentieth centuries understood that a nation's best defense lay in power projection, and only a strong navy could accomplish this. Geostrategists often talk of nations going through three phases of naval development. The first is called "brown water," where a nation can patrol its inland waterways and river systems. One can think of being able to patrol the Mississippi, Ohio, Wisconsin, and Colorado Rivers, plus the Great Lakes. The second is "green water," where that same nation has evolved to protect its coastline and immediate vicinity. In this scenario, America was able to patrol the Chesapeake and San Francisco Bays and the Gulf of Mexico. Finally, few nations reach "blue water," where they are oceangoing, global, going far beyond their own horizon. This is the America that projects power deep into the Atlantic and Pacific Oceans, the Mediterranean and the Red Seas, and so on. Space power advocates warn that obsession with using space only to change terrestrial battles is brown water thinking, focusing on LEO and GEO is green water, and those that understand the expansion of power into cislunar and beyond are blue water. Therefore, space power advocates often refer to themselves as the space "blue water school" (Goswami and Garretson 2020). Many use the color water terms frequently and classify focus on LEO and below as brown water (Worden 2021). They are concerned that the powers that be possess "a brown water perspective, envisioning a tiny Space Force with tiny budgets that merely operates uninhabited satellites in orbit around the Earth to support terrestrial warfighters" (C. Smith 2020).

As much as I agree with the sentiments of the blue water school, it would be better to break free from the bonds of the earthbound Navy and embrace the term "black water" to describe American national security into the high frontier. Further, the brown/green/blue debates may be counterproductive as our ultimate goal is not space power for space power's sake but the prospering and power of American civilization and culture. Unfortunately, some of the most gifted and forward-thinking space professionals' mistake is that their obsession with space leaves earthbound matters tangential. They are so excited about the grand issues of space-based energy, asteroid mining, lunar colonies, and even conflict in space that they forget that our ultimate goal is not space but American primacy and the American-led alliance system. This only makes sense as long as we preserve natural rights and natural law.

The time for thinking of space in purely reactive and defensive terms needs to end (Carlson 2020). Even the American people's focus must shift from thinking only in terms of science and exploration toward strategy and economics. For example, NASA has been accused of being in the intellectual doldrums (Impey 2016). The once shining star of all space programs has been stifled by old thinking and bureaucracy. Over the decades, Space navalists alter the Mahanian virtuous cycle and apply it to space.

Mahan's principles of the virtuous cycle illustrate that a nation's propensity for economic activity creates the need to transport goods on the sea. This merchant activity needs naval protection, which profits the nation, its business, and military power. In space, this can lead to phases of development from exploration to expansion, to exploitation, and even exclusion (of rival powers) (Carlson 2020).

An area of "navalist" space thinking that is most understood and therefore heartily criticized is when opponents of American space power argue that the parallels don't make sense since space has no "geography." Nothing could be further from the truth. Space is indeed a cold vacuum, harsh to humans in every way. It is also true that there are no space capital cities to take, no beachhead to storm, and no port to defend (yet). However, astro-geographic and astrographic areas are critical for both economic and security concerns. Mahan talked of the oceans as the wide commons with six fundamental elements of sea power: geographical position, physical conformation, the extent of territory, size of the population, character of the people, and character of the government (Mahan 1940). Many space power advocates illustrate the need to understand space in the same terms (Dolman 2002, 71). One needs to start thinking about space terrain such as gravity wells, Hohmann transfer points, orbits, and space choke points in LEO and GEO. Lagrange points, where gravitational and centrifugal forces are in balance, and places to give us strategic pause such as the Van Allen radiation belts and Kordylewski clouds, are of further interest for their strategic value. As such, there will be the advent of transportation and commercial routes that "develop specific pathways of the heaviest traffic" (Dolman 2002, 29). These will be based on finding the most efficient fuel conservation, security, and environmental safety routes (Carlson 2020). Naturally, these will also be the areas of contest and potential war, not unlike the straits of Gibraltar or Fulda Gap.

## SPACE FACTIONS

Different camps of thought dominate space policy. One camp consists of the scientists, technologists, and engineers who often have little interest in the political–strategic equation and, in a few cases, work against it. They often see space exploration only in the context of exploration for exploration's sake. A second camp consists of the traditional military, suspicious of ideas such as space domination and the need for a separate service at all. Yet another camp features the political class, who may understand the immediate value of the space program but fail to prioritize the right programs. Lastly, there is a camp that includes "NewSpace" advocates. They see the immense possibilities in space for economics, entrepreneurship, and colonization. Unfortunately, they also tend to be full of ranks of people who are not aware of geopolitics and astropolitics. Rarely has anyone articulated where the United States needs to be in 5, 10, 50, or

100 years—and beyond—to ensure it is the premier spacefaring nation. The United States cannot prosper as a spacefaring nation where there is a division between those that advocate astronauts and exploration and those that want to advance national space power.

Another good way to look at the various space factions is to analyze the visions of space. Three views are often highlighted. The first of these is known as Von Braunian (named after the German who had developed the V-2 rocket for Nazi Germany and then was recruited by the United States after the war). This vision is based on state control of space, primarily for military purposes and nationalist goals. Many would argue this was NASA during the Cold War. This school of thought is often referred to, somewhat derisively, as "flags and footprints." The second, Saganite (named after popular astronomer and cosmologist Carl Sagan), centers on noninterference and scientific discovery. This is one where there is a limit on human activity where one "looks but does not touch." Finally, the O'Neillian view (named for Gerry O'Neill, the famous physicist and space activist) emphasizes the need to go to space for the survival of humankind, growth, freedom, prosperity, and curiosity. If left to its own devices, this three-way fragmented view will lead to disaster and chaos.

It should be noted that not all space power advocates align. Although most see the blue water future (though again, I choose black water), some are still constrained by the problem of presentism. Bleddyn Bowen's 2020 book *War in Space* is an excellent example of someone who clearly sees the critical nature of space power but is unwilling to break the prison of present technology and geocentrism. First, he argues that one wages space warfare to command space to ensure the ability to use one's own most essential celestial lines of communication without major disruption or to deny an opponent the same ability. Bowen's second proposition stresses that space power consists primarily of infrastructure and remains connected to Earth because it serves to enhance "activities on Earth." The third proposition states that command of space does not equate to command of Earth. Bowen's fourth proposition centers on a kind of geography of space and Earth. He argues that command of space enables the manipulation of celestial lines of communication using "Newtonian-Keplerian and electromagnetic" chokepoints (Bowen, 2020, 89). The fifth proposition explains the implications of Earth's orbit being akin to a cosmic coastline in terms of strategic maneuvers. The sixth proposition states that space power functions inside "a geocentric mindset." His argument that space power is primarily based on a "proximate littoral environment of Earth orbit" (Bowen 2020, 105) proves the need for astropolitical and astrostrategic planners to go into the Space Futures Workshop mindset of thinking in 25-, 50-, and 75-year increments to continue American primacy.

A final view, a view that reflects the rudiments of the needs of the Space Force, was espoused by Lt. General Daniel Graham and is called by some the Grahamian view. It advocates that America seize the ultimate "high ground" (space) and use its superiority to dominate space for the realistic benefits of national power (Ziarnick 2015, 73–74). The USSF will need to address the "inherent weakness in the 'join' system of command, strategy, operations, and budgeting" (Dinerman 2021, 33).

The absolute priority is to fully integrate all these aspects of space into current and future national security and grand strategic thinking. The only way to accomplish any of this from a grand strategy perspective is to create a separate military service. A coherent space doctrine and strategy will be to integrate these camps, as much as possible, into a whole. There is room in space for many of these ideas, but only if there is general agreement on the foundation. This foundation is based on American space dominance and governance.

## CRITICISM

One of the main arguments against any of these proposals is that this will militarize space. The problem with this argument is that space is already militarized, and in some sectors, the Russians and the Chinese governments are ahead of us in both the military and civilian sectors. In order to dominate space, China and Russia are embarked on complementary space policies. They have in place versions of a militarized space branch, and more importantly, space dominance doctrines designed to dethrone the United States from its military and economic position.

Such studies include China's proposed work in space-based solar power (SBSP) and the testing of anti-satellite weapons and Russia's hypersonic missiles. The great powers realize that geopolitical imperative obeys no master. Interestingly, far from hiding this fact, Russia and China are blatantly open about it when we examine what China says about its Strategic Support Force, or Russia, about the Russian Space Forces. All of the great powers and some of the medium powers have recognized that space is the new measure of future national power (Dolman 2002).

Another argument, primarily in liberal political circles, is that we could mitigate all of this with international law. They cite the 1967 Outer Space Treaty and, in a few cases, a new, more restrictive treaty: PAROS, the Prevention of an Arms Race in Outer Space. This side does have an argument, in that the United States is the leading nation in the world and is the touchstone of international law. Thus, America should follow the mechanism for withdrawal from the Outer Space Treaty just as President Bush withdrew from the ABM Treaty and President Trump from the INF Treaty. These are treaties that our adversaries did not follow, which only resulted in reducing national security for the United States.

A final argument against this proposal centers on budgetary issues. Needless to say, there is going to be a cost to a new branch, although it is a cost dwarfed by the non-budgetary cost of America losing the strategic ground to its adversaries. However, money would also be saved if all the funds from the current uniformed civilian services with a small piece of overall space-related budgets were given to the USSF. It would also prevent the U.S. Air Force branch from raiding the space budget for their own purposes.

## BENEFITS

To achieve the grand strategic goals of space dominance, the United States must support President Trump's March 2018 call for America to be "First among the Stars." Trump not only wanted to reinvigorate human space exploration, as described in Policy Space Directive 1, but also to ultimately create a separate branch of the military (The White House 2017).

Again, if we fast-forward to 2020, we see the beginning of policy implementation.

National leadership recognizes the criticality of space to national security and prosperity. Space, including space security, is a top national priority with increasing resources to ensure continued U.S. leadership in this critical domain. . . . The creation of new space-focused organizations in DoD offers an historic opportunity to reform every aspect of our defense space enterprise. The USSF, the newest branch of the Armed Forces, will bring unity, focus, and advocacy to organizing, training, and equipping space forces. USSPACECOM, the newest combatant command, will bring additional operational focus to deterring threats and shaping the security environment in space. . . . New leadership and management for space acquisition has been established to unify the Department's space acquisition efforts into a streamlined structure for better integration and speed of delivery. . . . The United States has long maintained a robust and prolific arrangement of alliances and partnerships built on trust, common values, and shared national interests. . . . Commercial space activities have expanded significantly in both volume and diversity, resulting in new forms of commercial capabilities and services that leverage commoditized, off-the-shelf technologies and lower barriers for market entry. (Department of Defense 2020a)

It is incumbent on advocates of such a service to recognize the need to integrate the electorate into the debate. Voters need to be educated about the current value of space technology, such as GPS, weather forecasting, communication, and military surveillance, to name a few, and protections for generations to come. Without advocacy from the electorate, this effort is a nonstarter. Fortunately, the American people's spirit is built on going forward to the final frontier, and this can be an easy case to make.

Many analogies are made to creating the U.S. Air Force by the National Security Act of 1947 and the creation of the U.S. Space Force. A better analogy is that the National Security Act of 1947 was a complete shift to the professionalization of the national security system because America was at an unprecedented, existential crossroad. We are at that kind of crossroad today.

The benefits of a separate branch are myriad. It would be the tip of the spear for space-based missile defense, the only actual future for protecting Americans from the threat of nuclear annihilation. By removing it from any branch already in existence, the Space Force would have a single task, not burdened by the baggage and other missions vital to the U.S. Air Force and U.S. Navy. It would, therefore, integrate all elements of national security into space policy and space strategy without bureaucratic fragmentation and chaos. This would also include a Space Intelligence Service that aids the mission of this branch without adding to the data crush already suffered by the uniformed military intelligence organizations and CIA. Additionally, the Space Force would possess total responsibility and total accountability for America in space without hiding behind other priorities. It would also require its own training and promotion system not burdened by preconceived notions in the other services.

Finally, the creation of the Space Force would signal to America's adversaries the seriousness with which we take grand strategy beyond rhetoric. Therefore, the Space Force would be the foundation for American grand strategy in the twenty-first century and beyond.

This will not only be a military vision of space. In his groundbreaking book *Beyond Our Future in Space*, Chris Impey sets out a future timeline that could happen. It paints a picture that we have a vibrant commercial space industry created by routine space flight by 2035, the first viable colonies on Mars by 2045, routine space mining by 2065, and the first generation of humans that come of age without having been to Earth by 2115 (Impey 2016, 244).

## DEMOCRATIC VALUES

Space enthusiasts, NewSpace entrepreneurs, scientists, and space utopians often fail to understand a simple truth. They have been nurtured and protected by the American republic. The grand visions they seek, and the destinies they wish to fulfill, are only possible because of the supremacy of American power and values. Many American space power advocates promote the idea that America's moral superiority propels us to seize control of space (Dolman 2002). The idolized *Star Trek* offered a vision that was a victory for democratic values.

Science fiction aside, there is a fundamental point here. Power cannot tolerate a vacuum. Either space is dominated by authoritarian and totalitarian regimes bent on conquest, economic monopoly, and obedience, or space will embrace the liberal order of liberty under law and the American democratic values of life, liberty, and property. There is no alternative. America will need to follow a grand strategy devoted to its primacy to achieve this.

# CHAPTER 2

# History

Americans have dreamed of going to the stars for generations. The Apollo missions were thought to be the starting point for the United States to be a spacefaring people. Still, this dream drifted backstage as the political class allowed itself to be captured by the winds of popular culture and perceived expediency.

However, the history of the U.S. space program has been one of the fits and starts with an unclear trajectory. This trajectory was clear when America was put on the path of landing on the Moon. Once achieved, the program drifted without a clear aim, little of an overarching goal, with perhaps the exception of satellite dominance. *For All Mankind*, the TV series, poses an alternative history where the Soviets beat us to the Moon. This ignites an aggressive space race that keeps pushing the boundaries of technology and spaceflight, leading us to a path where we are 30 to 40 years ahead by the time we reach the 1980s and beyond. This proves a problematic point. America reacts well to threats, existential threats in particular. Existential threats cut away the fluff and nonsense of a society that dominates in peace and prosperity eras. However, we don't always want to live under such a sword of Damocles. This will be one of the complex issues for space power advocates to deal with, but dealing with it must be a priority. This pushed the U.S. space program inward-looking, viewing LEO and GEO's barriers as unbreakable, the province only of probes and landers. This has created a vacuum where America has come very close to making the mistakes of past empires because of shortsightedness and even negligence.

This lack of political foresight was clear in Athens during the Peloponnesian wars and the Roman Empire in the fifth century, China in the nineteenth century, France, and most of Europe in the 20th century.

Discussions during the Space Futures 2040 meetings included questioning the assumptions made about the level of technology we will possess and how that will translate into today's terms. I proposed thinking along the following lines: Imagine you were a naval professional in 1900 attempting to strategize about 1940. Who would be seen as more reasonable? The man who said that the 1940 war would be won by heavily armored dreadnoughts or the one who envisioned a ship where warplanes would take off and return to? The first realistic concept of the aircraft carrier did not occur until 1909, the first fundamental test of the idea not until 1911, and the first working basic carrier not until 1914.

In each case, the problem was that national leaders failed to appreciate the technological and strategic advances that made their political and military doctrines obsolete. For eons, across military history, myopic leaders have concluded the following:

- Fixed fortifications could withstand gunpowder.
- Longbow arrows could not fell heavily armored knights.
- Wooden ships could stand against armored battleships.
- French troops stationed on the Maginot line could defend France against a German invasion.

As a result, history is littered with the bodies of soldiers whose political leaders lacked foresight and imagination.

In 1897, the American naval officer and historian Alfred Thayer Mahan warned of the consequences of ignoring significant advances in military tactics and armaments in his book *The Interest of America in Sea Power, Present and Future*:

That the United States Navy within the last dozen years should have been recast almost wholly, upon more modern lines, is not, in itself alone, a fact that should cause comment, or give rise to questions about its future career or sphere of action. If this country needs, or ever shall need, a navy at all, indisputably in 1883 the hour had come when the time-worn hulks of that day, mostly the honored but superannuated survivors of the civil war, should drop out of the ranks, submit to well-earned retirement or inevitable dissolution, and allow their places to be taken by other vessels, capable of performing the duties to which they themselves were no longer adequate.

It is therefore unlikely that there underlay this re-creation of the navy—for such in truth it was—any more recondite cause than the urgent necessity of possessing tools wholly fit for the work which war-ships are called upon to do. The thing had to be done, if the national fleet was to be other than an impotent parody of naval force, a costly effigy of straw. But, concurrently with the process of rebuilding, there has been concentrated upon the development of the new service a degree of attention, greater than can be attributed even to the voracious curiosity of this age of newsmongering and of interviewers. This attention in some quarters is undisguisedly reluctant and hostile, in others not only friendly but expectant, in both cases betraying a latent impression that there is, between the appearance of the new-comer and the era upon which we now are entering, something in common. If such coincidence there be, however, it is indicative not of a deliberate purpose, but of a commencing change of conditions, economic and political, throughout the world, with which sea power, in the broad sense of the phrase, will be associated closely; not, indeed, as the cause, nor even chiefly as a result, but rather as the leading characteristic of activities which shall cease to be mainly internal, and shall occupy themselves with the wider interests that concern the relations of states to the world at large. And it is just at this point that the opposing lines of feeling divide. Those who hold that our political interests are confined to matters within our own borders, and are unwilling to admit that circumstances may compel us in the future to political action without them, look with dislike and suspicion upon the growth of a body whose very existence indicates that nations have international duties as well as international rights, and that international complications will arise from which we can no more escape than the states which have preceded us in history, or those contemporary with us. Others, on the contrary, regarding the conditions and signs of these times, and the extra-territorial activities in which foreign states have embarked so restlessly and widely, feel that the nation, however greatly against its wish, may become involved in controversies not unlike those which in the middle of the century caused very serious friction, but which the generation that saw the century open would have thought too remote for its concern, and certainly wholly beyond its power to influence. (republished, Mahan 1940)

As stated in chapter 1, Mahan predicted that a nation that invested in advances in sea power would dominate the globe. His prediction was ultimately demonstrated by the strength and power of the aircraft carrier during the Pacific campaign in World War II. However, those days are numbered, as were the days of the now obsolete battleship.

The United States is at a similar crossroads now. This crossroads presents us with decisions on whether or not we will lead in space, advance medicine through space technology, and have the capability to dominate the next battlefield.

### SPACE DREAMS AND SCIENCE FICTION

In what can only be described as an odd set of circumstances, space power and fiction have merged into their own synthesis, feeding off one another in a complex web of interplay and back-and-forth. In other words, the history of the space program, and the creation of the Space Force, are as much a product of actual history as the history of American-dominated science fiction. Therefore, any explanation of space strategy without an homage made to said fiction would be incomplete and dishonest. Admittedly, some in NASA and the military want to run as far away from the science fiction (especially *Star Trek*) model as their warp drive can take them. However, the visionaries one meets seem to share their adoration and even mimicry of the genre to propel them. This ranges from Elon Musk, using the video game *Mass Effect* to model his spacesuits, to Space Force officers wishing they were training at the Starfleet Academy.

In 2021 we observed the 55th anniversary of *Star Trek*. One expected a panoply of odes, eulogies, parodies, and parallels. Space Force has already been accused of copying Star Fleet's delta, whereas *Star Trek* copied the delta from the Army Air Force and NASA.

This is an excellent example of the problem that critics have of Space Force, albeit cosmetic. They believe that linking the U.S. Space Force to *Star Trek* will somehow discredit the organization. Perhaps they are trying to take a page from President Reagan's Strategic Defense Initiative attacks. But instead of debating its many merits, they thought they would castigate it with the moniker "Star Wars."

We can pause here and offer a lesson in strategy and tactics. If you want to discredit something associated with space in the American public's eyes, especially potential recruits, don't use the two most fantastic visions of space and space opera to do it. One would have loved to be in the meetings where someone voiced their proposal and said, "We don't want Space Force, so you need to convince all those children that if they join, they will be going warp speed and wielding lightsabers." If your demographic is people with no vision and imagination, you have yourself a winner.

As earlier stated, William Shatner's article "William Shatner Wants to Know: What the Heck Is Wrong with You, Space Force?" (Shatner 2020) created a stir as he advocated naval instead of army and air force rank by using science fiction standards. However, his more profound argument revolved around the need for heroes in the public mind, and this would best be done by linking Space Force with science fiction like *Star Trek*.

*Star Trek* offered a vision that was a victory for democratic values. It served and continues to serve as a foil to the antihero dystopia that passes for much futurism today. *Star Trek* exhibited the absolute nature of American values by recoiling at the horror of genocide ("The Conscience of the King"), harpooning futuristic tyrants ("The Apple"), and hippie culture ("The Way of Eden").

More importantly for Americans is that *Star Trek* represented an American vision of the future. This is not merely a representation of American patriotism but the universal values America champions. It ranged from the cosmetic where Capt. James T. Kirk was from Riverside, Iowa, to *Star Trek* promoting the importance of liberty, right reason, frontier spirit, and human rights and human dignity.

Star Fleet played the role of a futuristic military and exploration mission. This was akin to the American Army and Navy's nineteenth-century exploits and was the sword and shield of these values. Star Fleet promoted a neo-manifest destiny broadening Thomas Jefferson's "Empire of

Liberty." Adversaries like the Klingons and Romulans are totalitarian and authoritarian dictatorships bent on destruction and conquest.

The episode that expresses all of this is "The Omega Glory," where the USS *Enterprise*'s landing party finds itself thrust into a planetary war between the Yangs (Yankees) and the Kohms (Communists). The Yankees eventually defeat the Communists, and Kirk discovers that their worship words are the American Pledge of Allegiance and the U.S. Constitution. Kirk states in the famous ending speech: "Among my people, we carry many such words as this from many lands, many worlds. Many are equally good and are as well respected, but wherever we have gone, no words have said this thing of importance in quite this way. Look at these three words written larger than the rest, with a special pride never written before or since.

"We the people of the United States, in order to form a more perfect union, establish justice, ensure domestic tranquility, provide for the common defense, promote the general welfare, and secure the blessings of liberty to ourselves and our posterity, do ordain and establish this Constitution for the United States of America. These words and the words that follow were not written only for the Yangs, but for the Kohms as well!"

I have read countless reviews of this episode from pseudo-intellectual critics who decry it as "the worst." They complain that it is overly patriotic, racist, and impossible for a planet to develop such a parallel conflict.

Kirk, whose hero was Abraham Lincoln, is a good starting point for dismissing this episode's critics and a *Star Trek* link to the American future. Lincoln, whose classical conservative roots stressed the Declaration of Independence's universality, was founded under the fatherhood of God and under God's natural law. Lincoln understood that these values transcended time and space and were literally universal. It is precisely the point that a space dominated by Western powers will be a space dominated by universal values based on the natural laws of life, liberty, and property.

Perhaps equal in impact to the space program's history and its future is the *Star Wars* franchise. This may illustrate a darker side to the power of the genre. On May 25, 1977, the original *Star Wars* movie, A New Hope, made its debut. It immediately had an impact that is hard to measure, especially on the generation that would, unfortunately, be called "X," itself a seeming sci-fi moniker.

Watching that film and the subsequent two sequels, there was no question whom one would root for. Everyone wanted the forces of the Republic to win, over the empire. These two forces are the science fiction versions of the American revolutionaries. But, on the other hand, the dark side, represented best by Darth Vader, was no British officer serving George III, but a minion of the likes of Hitler or Stalin.

The Force has a good side and an evil side. The dark side represents violence, torture, repression, death—evil. The light side is attracted to order based on morality, ethics, courage, and justice. The dark side is drawn to order based on power, betrayal, and greed.

It is astounding to witness the rise in popularity of the dark side among the makers of the *Star Wars* movies and merchandizing offshoots. There is a reason that the dark side's foot soldiers are called "storm troopers" and that young Darth Vader commits mass murder of children in Episode 3, *Revenge of the Sith*. This is a disturbing trend whose roots run far deeper than movie criticism.

Disney stores and a host of department stores promote equality, if not more partiality given to the evil side, stocking toys and games that enhance the "coolness" of Vader and the storm troopers.

They give equal time to toys and costumes of the dark side; we have video game companies like Electronic Arts, whose blockbuster game, *Star Wars Battlefront II*, forces players to take the role of a dark side champion

As a father and a *Star Wars* fan who has watched this trend over time, the appropriate reaction can only be a reprise of the famous phrase, uttered by Obi-Wan Kenobi: "I felt a great disturbance in the Force as if millions of voices suddenly cried out in terror and were suddenly silenced." It cannot be accurately quantified, not in dollars and cents, and not in psychological studies. The very fact that any sane person could venerate Darth Vader (before he recants at death), Emperor Palpatine (who gives Caligula a run for his money), or Grand Moff Tarkin (who kills every man, woman, and child on an entire planet, regardless of the brilliant performance by Peter Cushing,) is beyond comprehension.

The primary cause of all this is an attempt by the political left to force moral relativism down the throats of every American, combined with a dogma called Red Puritanism, in which there are no absolute goods except for a laundry list they have deemed worthy: multiculturalism, tolerance, atheism, socialist realism, skepticism, activist science, anti-Western ideology (extra piety points for being anti-American), and collective White guilt. There are no immutable goals except for those prescribed by their dogma: ending White privilege, destruction of conservatism, the cult of victimization, a reduction of American military power, and the glorification of anything that shocks.

One can anticipate four criticisms of this analysis. The first is that I am overreacting to a pop culture science fiction movie. Americans in this camp should realize the power of epics, storytelling, legends, myths, and the use of the English language. *Star Wars* changed all of this from 1977 onward and continues today with the release of the newest film.

The second is that only a political scientist or historian would read so deeply into a movie. Except, *Star Wars* franchise has millions of followers, generates billions of dollars, and consumes untold hours of conversations and, dare one say, heated arguments. It has the lasting power of a story like *The Lord of the Rings*, or anything by Shakespeare, on the Western psyche. It is debated and discussed more today than any other fictional story among large sectors of the population, perhaps even more than any current debate or issue in politics today.

The third is the idea that moralizing against *Star Wars* is antithetical to free choice because no one is forced to root for Team Vader. This is reasonable if you embrace the relativistic argument that one can be as easily attracted to Nazi Germany, Soviet Russia, or militarist Japan as to America in World War II or the Cold War. Only as a historical parallel can this argument make sense, but it falls under its own intellectual weight.

Finally, one can argue over which ideas filmmaker George Lucas attempts to promote. Whatever his politics, they are not as relevant as the impact of the films' cultural phenomena.

One does not need to lapse into hysteria to raise these concerns, nor is there any solution that could or should be mandated by the government or the entertainment industry. However, that should not stop those who prefer the virtues of the Force to the dark side from asking Americans to reflect on the effects of their silver screen infatuation.

One could go on and on about the impact of fiction on the ideas of space power, and the list would be endless, including titanic works like Stanley Kubrick's film *2001: A Space Odyssey* and Robert Heinlein's novel *Starship Troopers*.

This is also not to discount the immeasurable impact of nonfiction works such as *The High Frontier* by Gerard O'Neill and space visionaries such as Jeff Bezos who would admit that this work, and this person, inspired them to climb the space ladder.

## MILITARIZATION OF SPACE AND FACTIONAL FIGHTING

No treatment of space power and strategy would be complete without addressing the issue of the militarization of space. In essence, this is an illusion. Space programs were driven from the beginning by military and security issues. The most blatant and overwritten example was the American reaction to Sputnik. It is further a mirage because America's adversaries have militarized and will militarize space regardless of any action by the United States.

American partisan politics is in full gear here. The political battle is "often portrayed as a fight pitting idealistic arms control enthusiasts who oppose all weapons against warmongering militarists who never saw a weapon they did not like" (Mueller 2010). Many left-wing thinkers argue that the American military is concerned with unproven "shibboleths" that will create the conditions for space war (Johnson-Freese 2016). Karl Mueller's *Totem and Taboo* describes three sanctuaries (anti-militarization) perspectives: "idealists" who want space to be unpolluted by the military, "internationalists" who believe that weapons create international instability, and an odd group of "nationalists" who believe space weapons threaten U.S. primacy. He also identifies three pro-weaponization camps as "space racers" who argue that the United States has no choice once it recognizes other states are militarizing space; "space controllers," dominated by many in the U.S. armed forces who see the military in space as necessary, practical, but politically costly; and "space hegemonists," those, like this author, who believe that space is the critical arena, the critical battlefield, and a guarantee of U.S. primacy (Mueller 2010). An attempt at a middle ground can also be found by those advocating what best would be a pause, perhaps akin to those in the 1980s advocating a nuclear weapons freeze. These advocates think that the militarization of space should be "delayed indefinitely." It is the recognition that space will be weaponized and that platforms like satellites need to be defended without overt provocation based on building offensive weaponry (O'Hanlon 2004). One has to question any strategy that recognizes the inevitable seriously but wishes to be second or last to use it.

## COLD WAR AND SPACE

One of the clearest examples of how great power conflict will continue in space is the origin of the space program, not only in the United States but abroad. The space program was a child of the Cold War between the United States and the Soviet Union. Its genesis was the various ballistic missile programs, initially a Soviet reaction to U.S. manned bomber superiority (Dolman 2002, 93). This then morphed into rivalry and races in robotics, probes, exploration, satellites, orbits, to its apotheosis, the race for the Moon. Pre-Cold War pioneers were pivotal. They included men like Robert Goddard, father of the liquid-fueled rocket; Wernher von Braun, whose Second World War career led him to develop the A4 and V2 rocket and then move on to a center of gravity at NASA; and Sergei Korolev of the USSR, who also used his WWII experience to develop the first Sputnik and put the first man in space, Yuri Gagarin. The space race would kick in when both the USA and USSR realized the need for ballistic missiles (Impey 2016, 36). It quickly became apparent that space would hold the key to three military arenas: communications, navigation, and surveillance (Dinerman 2021).

Obviously, the Second World War was the catalyst to the space program led by Britain's development of radar, German rocket technology, American computer advances, and the American atomic bomb (Dolman 2002). Again, this is a salient point that the space program was born of blood and war and cannot shake this association, even if it wanted to. It would be the Eisenhower administration that would govern the crossroads.

Both presidents Truman and Eisenhower feared the creation of a militarized state. Eisenhower's science team did not see the value of space and thought space could be "traded" for various peace overtures (Ziarnick 2015) This attitude of space's lack of importance frustrated development of a comprehensive space policy (Cooper 2021). However, Sputnik in 1957 not only scared the Eisenhower administration out of their lethargy but led to the creation of NASA. It would also set the stage for American dominance as Sputnik was renamed "Khrushchev's boomerang" by John Foster Dulles (McDougall 1997). The space race was on, culminating with the Mercury, Gemini, and Apollo missions. The missteps here, as President Reagan's dynamo of the Strategic Defense Initiative, Hank Cooper, explained, created many of today's problems. "The transfer of power from the Eisenhower to Kennedy administrations more completely converted these concerns into bifurcating our space programs into 'peaceful' space activities led by NASA and 'intelligence' activities led by the supersecret organization, the NRO embedded in the Air Force, in conjunction with our ICBM/SLBM programs. When Gen. Bennie Schriever (the father of USAF Space) tried to combine the two worlds, he was throttled back by the 'powers that be'" (Cooper 2021). The vaunted Outer Space Treaty of 1967 was not an example of pacifism and diplomatic nicety but of realism and rivalry (Dolman 2002). The technocrats took over during the Kennedy and Johnson administrations and ultimately melded image and reality (McDougall 1997), the apex of which was President Kennedy's speech at Rice Stadium.

"We choose to go to the Moon. We choose to go to the Moon in this decade and do the other things, not because they are easy, but because they are hard, because that goal will serve to organize and measure the best of our energies and skills, because that challenge is one that we are willing to accept, one we are unwilling to postpone, and one which we intend to win, and the others, too" (Kennedy 1962).

However, voices in the wilderness knew that this trajectory could not end with a moon landing as Generals Bernard Schriever and Thomas Power predicted (Ziarnick 2015, 86).

**SPACE WEAPONS**

Space weapons have captured the imagination for generations. The future will see the development and deployment of space weapons to be used in space and terrestrial warfare. Still, the focus is on weapons that can attack platforms in orbit like satellites (anti-satellite weapons) and those weapons that could be placed in space to attack targets on Earth. In nightmare scenarios, adversarial space powers use WWII-era strategic bombing doctrines to destroy cities and our economic and industrial base (Ziarnick 2015). As many advocates see space weapons as inevitable and necessary, some caution that they are not a "silver bullet" for our defense needs (Hitchens 2016).

Just as there is a continuum regarding the militarization of space broadly, there are specific continuums for space weaponization. This continuum can be illustrated along parallel lines ranging from no military use, support for the Earth-based military, acknowledging the necessity of space for the Earth-based military, and finally advocates of weapons in space. This can also be expressed by having nothing in space, only fighting in space, and using space weapons against terrestrial targets (Mueller 2010). The actual weapons continuum envisions energy weapons and weapons of mass destruction (WMD) at the extreme end. There is an ultimately inside-baseball debate among space professionals about what a weapon of mass destruction is. I have been part of such debates and have witnessed the back and forth. International relations experts usually qualify WMD weapons as nuclear, chemical, biological, radiological, and cyber. There is no magical number of casualties. The term "mass casualties" is used. Thus, it is straightforward that if you drop a 50-megaton nuclear bomb on New Delhi, that is clearly a WMD. However, a radiological bomb that ultimately kills thousands of people, primarily in its immediate aftereffects and chaos, would also be a WMD. If a space weapon inflicted mass casualties in the same way as any of these, it would also be a WMD.

A critical system that had huge political ramifications was space-based missile defense, pivotal to President Reagan's Strategic Defense Initiative (SDI) and mocked by some on the left as "Star Wars." One of my first experiences in Washington was attending a debate between the senior foreign policy advisors to Senator Jesse Helms and Joe Biden. I sat in the small hearing room and listened as both men displayed their acumen as surrogates regarding missile defense and President Reagan's SDI legacy. It astonished me that Biden's man had so little understanding of *realpolitik* and, in particular, the goals of our adversaries. The back-and-forth continued until the Biden representative retreated into the old canard that the SDI vision could not be accomplished regardless of the political issues because of the problem of technology. During the Q&A, I distinguished myself as a member of the minority in the audience by asking the following question, which I still ask today: "In the end, your argument is about a lack of technology and innovation, but that is not your real problem. If we had the technology today, would you still be against it? Is your real problem a disdain for American primacy?" In the typical myopic view of those who view MAD (mutually assured destruction) as a religion, he admitted as much. This experience proved telling, as countless times anti–space power advocates would claim that the science and technology did not allow for its creation, when in reality, they did not like the policy or the system being developed. I am one of those who believe that technology will follow mission and policy. Create the strategic conditions you want and ask for the technology and engineering to follow. This is how America reacted to every major conflict it has ever fought successfully.

If it had been developed as it was planned, SDI would have advanced American space power exponentially with programs like laser and plasma weapons, particle beams, and the ballistic missile defense system Brilliant Pebbles. Space-based missile defense is still the only true path to a solid defense against ballistic missiles. The Outer Space Treaty (OST) forbids many of these weapons. "States shall not place nuclear weapons or other weapons of mass destruction in orbit or on celestial bodies or station them in outer space in any other manner" (United Nations 1967). However, the USSR developed its fractional orbital bombardment system even after the OST. As noted earlier, there is little consensus aside from nuclear weapons on what would constitute a weapon of mass destruction. Finally, IR realism (the concept that nations act out of pure self-interest and the competition for power) would dictate that any attempt to use the OST to enforce such a ban would be as meaningful as when the League of Nations attempted to stop German and Japanese aggression. A weapon system in this same category is non-nuclear. Still, it would have the impact of a WMD under the moniker of "Rods from God," where tungsten rods are launched from orbit to the Earth creating a massive kinetic effect equal to a nuclear strike. Finally, an electromagnetic mass driver can be developed that could be utilized as both a launch and propulsion system and a rail gun weapon.

## THREE CASE STUDIES IN FAILURE

Three historical case studies offer us a window into small-minded narrow thinking that prevented the use of these ideas to propel the United States into space power dominance. These were the development of the ICBM and the failure to pursue the Horizon and Orion projects.

Both the USA and the USSR dovetailed off the German rocket developments of WWII. Yet it was the Soviets who moved ahead of the United States in ICBM and satellite development as early as 1953, while the USA cut off funding the program by 1948. The reasons echo today: There were concerns about cost, there was an assumption of the "permanent dominance" of American air superiority, discussions inside the USAF were dominated by such fighter and bomber pilots, and the American scientific community was generally skeptical (McDougall 1997, 96–97).

The Horizon project was based on a 1959 U.S. Army study, directed by Lt. General Arthur Trudeau, to construct a lunar military base, on the Moon. "There is a requirement for a manned military outpost on the Moon. The lunar outpost is required to develop and protect potential United States interests on the Moon; to develop techniques in moon-based surveillance of the earth and space, in communications relay, and operations on the surface of the Moon; to serve as a base for exploration of the Moon, for further exploration into space and for military operations on the Moon if required; and to support scientific investigations on the Moon" (United States Army 1959). Had Horizon been implemented, not only would the Moon landing have occurred earlier, but American sovereignty of the Moon would have been established before the Outer Space Treaty. One of the most telling sections of the plan states, "the presently contemplated earth-based tracking and control network will be inadequate for the deep space operations contemplated. Military communications may be greatly improved by the use of a moon-based relay station. The employment of moon-based weapons systems against earth or space targets may prove to be feasible and desirable" (United States Army 1959).

Finally, the idea behind the Orion project was to create a "deep space force" nuclear-powered space fleet (Ziarnick 2015). Initially the brainchild of physicist Freeman Dyson, the concept revolved around building a military space fleet powered by nuclear explosions. The 1959 *Implication of Orion Vehicle* report envisioned the creation of a military force of Orion starships that would operate in LEO and beyond to prevent the "disastrous consequences of an enemy first." This 20-ship fleet would have a crew complement of 20 to 30 people per ship, all containing artificial gravity, and be a self-sustaining American military mobile space base (G. Dyson 2003). Dyson himself claims that the number one hurdle to developing this was the "Von Braun brigades" of heavy chemical rockets (Dyson 2016). Had the United States pursued Orion, not only would we already have space superiority, but we would have already established long-range interplanetary travel.

Here we have three massive historical failures of vision and imagination. The proper decisions here would have guaranteed American national security for the future. Instead, they all share the same problems. These are the intransigence of the status-quo defense establishment, the naysaying of the science and technology elite, and the lack of vision of the political class.

**THE TWENTIETH CENTURY**

The march toward space power continued through the twentieth century. In March 1961, General Bernard Schriever and Asst. Secretary Trevor Gardner issued a report recommending the development of manned space flight, space weapons, reconnaissance, space stations, and lunar landings (McDougall 1997). The report warned of Soviet space supremacy and the need for a crash space program. Ultimately, the powers that be viewed the conclusions as too controversial.

The Apollo program and ultimate landing in 1969 seemed to predict an American future for space, a spacefaring people that would conquer the stars in the name of God, glory, and democracy. The American flag planted on the lunar surface became the single most distinguished icon of space achievement. Yet it was the end. Apollo became the victim of its own success, eventually to be "discarded, dismantled, distorted" (McDougall 1997, preface). Just as the Schreiver/Gardner report criticized the Eisenhower administration for attempting to divide the space program between civilian and military, the Apollo program seemed to confirm such an artificial bifurcation in many minds. The Eisenhower, Kennedy, and Johnson administrations utilized the U.S. vs. USSR satellite competition for enhancing alliance building and collaboration (Grosselin 2021). We, the United States, had won. All we had to do now was rest on our laurels and remind everyone that only one flag was on the Moon. This was one of the worst decisions in American history.

There were fits, starts, spasms, going forward and going backward. Typical of the Carter years was double-minded thinking. In 1978, President Carter's NSC 37 encouraged the commercialization of space, while maintaining the need for civilian and scientific exploration and advancing the "interests of the United States" (The White House 1978). This did not lead to any fundamental change. In the late twentieth century, Speaker Newt Gingrich's 1981 bill, the National Space and Aeronautics Policy Act, envisioned colonies and set forth ideas about space colonial government. This was a real vision before its time. It was mocked by many as fantastical. No serious national security expert is mocking it now.

President Reagan's NSDD 42 was slightly bolder about U.S. security and superiority but was primarily focused on the space shuttle (The White House 1982). In 1985, the United States took robust action to create U.S. Space Command, a unified combatant command for military operations in outer space. One year later, the Reagan administration became more forthright: "The directive further states that the United States will conduct those activities in space necessary to national defense. Space activities will contribute to national security objectives by (1) deterring, or if necessary, defending against enemy attack; (2) assuring that forces of hostile nations cannot prevent our own use of space; (3) negating, if necessary, hostile space systems; and (4) enhancing operations of the United States and Allied forces" (The White House 1988).

## TWENTY-FIRST CENTURY

It was not until the George W. Bush administration that a realistic merging of national security and space occurred. First was what has been called the Rumsfeld Commission in 2001, which called for the transformation of the U.S. military for space. "Use the nation's potential in space to support its domestic, economic, diplomatic, and national security objectives. Develop and deploy the means to deter and defend against hostile acts directed at U.S. space assets and against the uses of space hostile to U.S. interests.... Employ space systems to help speed the transformation of the U.S. military into a modern force able to deter and defend against evolving threats directed at the U.S. homeland, its forward-deployed forces, allies and interests abroad and in space" (Commission to Assess United States National Security Space Management and Organization 2001). This report clearly endorses American space superiority. Even more critical, the report called for creating a Space Corps. It can be argued that America took a step back when Space Command was reduced to the U.S. Air Force Space Command in 2002. Space power opponents used the national security shift created by 9/11 to eliminate the unified command (Smith 2017). This was followed by President Bush's 2006 National Space Policy, which promoted national security in space and space professionals (United States 2006). The Obama-era space strategy recognized that space was "increasingly congested, contested, and competitive" (United States 2011). However, it used the word "encourage" a lot and did not mention how space needs to merge with overall national security and grand strategies. It also did not bode well that the Obama administration abolished the National Security Space Office (Dinerman 2021, 30).

Our contemporary period birthed an understanding of black water space navalists who understand what the secretary of defense noted when he clarified that space is already contested (Broad 2021). This period saw the proposal for a Space Guard, a Space Corps, and finally a Space Force. In 2016, Congressmen Jim Cooper and Mike Rogers offered legislation to create a Space Corps inside the Air Force, but the forces of the status quo stymied this. Rogers noted that none of the Air Force colonels up for general promotion were space professionals (Rogers 2017). The space navalists noted that "American space power stagnated under U.S. Air Force stewardship" (Smith 2017). In 2018, President Trump's National Space Strategy was the most courageous. It outlined a strategy of putting American interests first, understood the "dynamic interplay" between national security and civil and commercial space. It declared the need for American space preeminence. "The National Space Strategy protects our vital interest in space—to ensure unfettered access to and freedom to operate in space to advance America's security, economic prosperity, and scientific knowledge ... recognizes that our competitors and adversaries have turned space into a warfighting domain" (The White House 2018). Space navalists like Josh Carlson described the inevitable four phases of space development: exploration (going to a new domain), expansion (extending to control locations and resources), exploitation (getting new resources), and exclusion (protecting from hostile actors). These form a back-and-forth synthesis, creating space superiority and dominance conditions (Carlson 2020).

President Trump's National Space Council resurrected the idea of a Space Force and set the path to creating the independent USSF on February 19, 2019, and the reestablishment of U.S. Space Command in August of 2019.

We had reached our space Mahanian moment. It was a long and tortuous history, but ultimately America is finally back to making the correct choices that were missed during the Cold War.

# CHAPTER 3

# Grand Strategy and Space Dominance

As stated in chapter 1, grand strategy harnesses all elements of national power over a long arc of time ranging anywhere from 25 years to centuries. Grand strategy sees the horizon as the starting marker and moves beyond it. It requires forward-thinking and creativity that many policy-makers, pundits, and academics are sorely lacking. Finally, it requires equal parts knowledge of geopolitical history, future trends, and most important goals, benchmarks, objectives, and mission. Without this last part, the other elements are irrelevant and useless.

## RETURN OF HISTORY

We are witnessing the renewal of two historical debates put on hiatus in many ways—the first of which stems from the beginning of the age of exploration. The debates about risk and reward, the unknown, the daring, the tragic, and the opportune reignited as if it were 1492 all over again. At that one moment in world history, the great powers' decision over the Americas and Asia determined their destiny for 500 years. The second debate is quintessentially American and erupted in the late nineteenth century, especially in 1892 and the Spanish American War. There were many questions at this time about political, moral, and economic considerations as America was transformed from a regional to a great power and its navy from brown water to blue water. One of those questions was pure Americanism: "Does the Constitution follow the flag?" Again, we will face this question as America establishes bases and ultimately colonies on the Moon, Mars, and beyond. All of this is particularly true of space and astropolitics. The thinking required is more problematic as it no longer is bound by a terrestrial anchor. Space is more dangerous than the oceans and the territories more inhospitable. Equally, the opportunities are vaster, and the possibilities limitless, as long as America gets it right.

In my book *The National Security Doctrines of the American Presidency*, I identified nine primary markers to successful principles under the prime assumption that they obeyed the framework of grand strategy. These variables were: American exceptionalism; manifest destiny and expansion; the empire of liberty and democracy promotion; free trade, commerce, and markets; unilateralism; internationalism; the American way of war; geography, geostrategy, and geopolitics; and primacy. First, a doctrine ensures that the nation's strategic-level security interests must function within a comprehensive national security doctrine capable of resisting time fluctuations, shifting administrations, party lines, and personalities. In addition to safeguarding its citizens, land, and way of life, American national security includes the expectations of protecting individual freedoms and national values. In other words, these doctrines understood the grand strategic American trajectory. The dominant subject in American national strategy since 9/11 has been counterterrorism. Still, intelligence and counterterrorism polies are not national security policy, and national security policy is not national security strategy, which in turn is not grand strategy or national security doctrine. Third, nothing is more critical to the survival of American civilization than a coherent, competent, and robust national security doctrine that can provide a foundation for grand strategy. We will return to these in the final chapter. Other analysts have offered sympathetic corollaries to American strategic culture, such as the frontier spirit, a strong preference for a rules-based international system, and an overall view that antidemocratic regimes are inherent threats (Goswami and Garretson 2020, 153).

## LADDER OF DECISIONS

We can imagine a ladder of decisions where tactics are the first rung (Figure 3.1). Here we have decisions about specific plans to counter a particular terrorist act, launch a drone strike, or build a new aircraft carrier. Then we climb up to the next rung, where there is an operational strategy to decide where to place that aircraft carrier or how we use drones.

The third rung is national security policy, where we create plans for the duration of time a leader is in office, which may even include deciding to go to war. The fourth is national security strategy, encompassing decades of decision-making and choosing midterm to long-term goals. At the very top, the fifth and final rung is doctrine and grand strategy—the place where decisions stem from centuries of evolving yet organic choices based on a nation's history, culture, religion, and geography. If national security is done correctly, the decision *begins from the top rung and determines what you do all the way down the ladder.*

If the flow is bottom-up, we often witness disaster. Think of all the talk today about the use of drones in Afghanistan and Pakistan and whether to give arms to the Ukrainians. All of this is important, but how often do you hear someone step back and talk about this in the context of how this will help in 5s, 10, 50, and 100 years in the future?

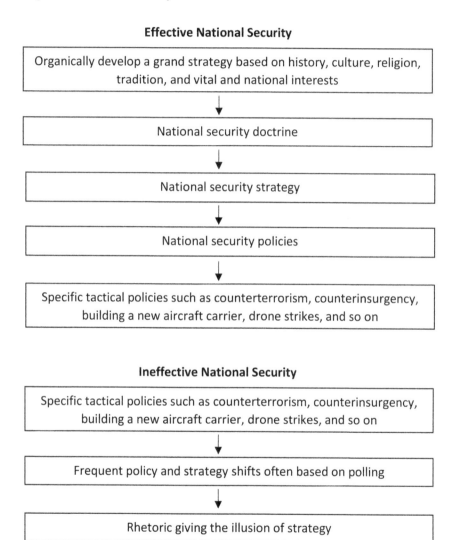

**Figure 3.1** Ladder of Decisions

Naturally, a country without a grand strategy is not only an airplane without a pilot but an airplane that will either crash and kill all aboard or be hijacked by a hostile power.

American grand strategy has always attempted to look to the past while acting for the future. Contemporary American grand strategy has tried to learn the lessons of the Cold War as that has been seen as an architecture that is clearly American. It is one where the Reagan years are often cited as being successful at combining the economic, the political, the ideological, the military, and the moral (Brands 2015). Grand strategy is not only based on military power, but the absence of military power makes any attempt at such a strategy futile. This military dimension illustrates again the need for grand strategy created by national security doctrines, which hold the reins of leadership instead of the reverse logic of allowing tactical considerations to dictate terms. The contemporary American military often seems hobbled by being "charged with 'winning' battles through 'losing' wars, and with 'winning' wars but failing to attain national security objectives by disregarding the lessons of history and duplicating past mistakes" (Collins 1973). Grand strategy must also be based on reality and not purely on unrealistic ambition (McDougall 2010). Grand strategy is often so deeply rooted in the history and substructure of the nation that policymakers and generals are not even aware they are serving it (Friedman 2009, 39). It should also be remembered that national character affects a nation's strategic decisions (Morgenthau 1992, 122–128). If a grand strategy is effective, it must be forward-looking and long-term (Hentz 2004). Realists will argue that grand strategy is focused merely on the security and safety of the nation. Hans Morgenthau divided these into a primary interest of security and secondary interests based on ideology, warning never to permit the secondary interests to overcome the primary (Morgenthau 1970).

Grand strategists are often divided among four schools—neo-isolationists, liberals, realists, and primacists (Cox and Stokes 2008, 19). However, instead of rehashing the debates among neo-isolationists, liberals, realists, and primacists, it is worth noting that the most successful national security doctrines never embraced this choice. Instead, they dealt with the time and space of their era and can transcend their time and space for the future (Gaddis 2005, 380). American grand strategy has always embraced the pragmatic and the ideological inheritance from the enlightenment (Hooker 2014). Grand strategies exist over large swaths of time: they combine a nation's historical imperatives with national character. We see this through historical examples such as Rome's determination to dominate Europe and the Mediterranean to spread civilization and law (see my book *The International Relations of the Bible*), and Great Britain's resolve to forestall a European continental hegemon to spread British mores, manners, and justice. Naturally, a considerable focus of American national security doctrines is to harness these imperatives and goals.

Grand strategy is the expression of doctrine in practice. These will be the exact same attributes America needs to achieve and maintain space dominance, and we will return to this in the final chapter.

## Space Power

Space power, as previously illustrated, involves the ability to utilize the area in and around space. It creates the terms of control and dominance. A basic working definition comes from Lt. Colonel David Lupton, where he includes certain characteristics: Space power is the ability of a nation to exploit the space environment in pursuit of national goals and purposes and consists of the entire astronautical capabilities of the nation (Lupton 1988, 15). Critical to our understanding of the space realm is that the geopolitics of the Earth will be reflected in space. Space will be the chessboard of the great powers as Earth has been for all of human history. It will be the marriage of geopolitics with the uniqueness of space that will produce astropolitics. The unique aspects of space terrain, resources, and astromechanical and physical limitations will be required to make this marriage understandable (Dolman 2002, 7–8). Some have divided this into a "grammar" or the collecting and linking of equipment, production, shipping, and colonies and a "logic" of application including economics, political, diplomatic, informational, and military (Ziarnick 2021). This combines the expansion of humans in space and development in the economic realm. A very holistic definition was created by James Oberg, encapsulated by six broad tenets. "1. Space operators should understand the advantages and limitations of operating in, to, and from space. 2. Spacepower should be prioritized and coordinated by a space professional with a global perspective. 3. Spacepower in a theater should be centrally controlled by a space professional. 4. Spacepower is flexible and versatile. 5. Spacepower is best used to achieve effects in and from space that capitalize on its unique advantages. 6. Spacepower can support or be supported by terrestrial operations or operate independently" (Oberg 1999).

Many space power advocates address the tragedy that American history is motivated by immediate threat. These aggressive catalysts, or strategic traumas, force policymakers to finally make the necessary changes in national security, but only during a period of emergency. These emergency points are well known to many readers. In the era of wooden ships and iron men, there were the Barbary pirates, causing America's first overseas war and proving the need for the navy; the burning of the Capitol; the explosion of the battleship *Maine*; the attack on Pearl Harbor; and of course, 9/11. Less aggressive but just as impactful would be the Sputnik program. The problem with this method of policy change is that engaging such change during a time of emergency contracts your options and is primarily reactive.

This mindset is problematic, as many space power advocates understand the need to always be on the strategic offensive. However, there is also a danger in thinking too militaristically, producing what Brent Ziarnick illustrates as being on the tactical offensive (focused on defeating Russia and China) but remaining on the strategic defensive. Being on the strategic offensive requires a focus on controlling space (Ziarnick 2021). Control is three-pronged and includes the scientific (knowledge), economic (wealth generated from space), and military (removing threats and protecting assets). Many cite the works of Admiral Wolfgang Wegener's three indices of sea power: strategic position, the fleet, and strategic will to the sea (Holmes and Yoshihara 2009). This is seen strategically as a virtuous cycle advocated by great thinkers such as Alfred Mahan, Sir Julian Corbett, and Theodore Roosevelt. A cycle was established where trade generates wealth, funding a navy that protects that trade, enhancing trade by creating stability, and ultimately creating maritime supremacy conditions (Figure 3.2). This is seen as easily extended into space where the Space Force protects assets, access, and people. This makes the conditions for space wealth, trade, and commerce, which creates greater resources for the Space Force, amplifying the cycle in turns.

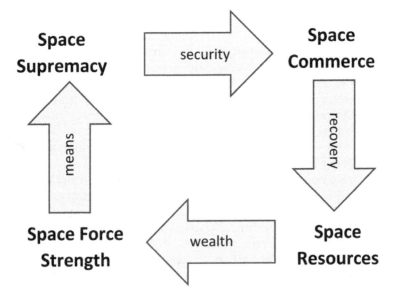

**Figure 3.2** Cycle of Support

## Space Strategy and Policy

The basic arguments for strategy and policy are here, but specific prescriptions will be in the final chapter.

### *Phase I: Strategy and Doctrine—Now to 10 Years Out*

Space Force will become the main instrument of American national security policy and, by extension, American foreign policy. Therefore, we need to fit space doctrine and strategy into the current 5 to 10 years and beyond 25 to 50 years. This was the thinking of the chief scientist of the U.S. Space Force, visionary Dr. Joel Mozer, and dynamo behind Space Futures workshops. Dr. Mozer's essential list of benchmarks is an excellent starting point.

- We must be able to protect and defend our military space assets starting now.
- We must be prepared to defend all U.S. assets and activities (commercial, private, government) within the next 10 years (as overall use of space increases).
- Our systems must be resilient to attack.
- Space-domain command and control must integrate seamlessly with air, land, sea domains.
- We must guard against technological surprise in space.
- We must provide cost-imposing technological dilemmas to our adversaries in space (Mozer 2021).

As General Raymond put it, "I think our big goal is to maintain that assured access and freedom to maneuver. I think that's critical to our nation. It's critical to how we are behaving in space. . . . And so for the next 5–10 years, on the space side of national security, we have to make a shift to treating space as a warfighting domain" (Raymond 2021). It is good to be reminded, as General Kehler stated, that "Space Force is a military service" (Kehler 2021). It cannot and should not try to do everything in space and focus only on the national security issues of space. This is still a galaxy-wide mission set. In this initial period, Space Force must be built, much from the ground up, and demonstrate effectiveness, capability, and success. It will need to expand beyond the initial parameter outlined in its original creation. America will need to "maintain a presence in the topography of the space domain" (Shaw 2021). Many space power advocates continue to warn that we need to shift to "blue water" thinking during this phase and believe that all space bureaucracy will be united under a Department of Space (C. Smith 2020). Regardless, this will be the time for building the structure and infrastructure of USSF and its initial doctrine and strategy. If this is not achieved correctly, Americans will not write the future. The United States initially needs to solve the problem of defense in orbit (J. Carlson 2021). This era will need to protect the initial human presence in space, such as search and rescue (Moloney 2021). This will also be the phase where we begin building support for the new economic revolution and economic development in space (Worden 2021). As General Shaw noted, the Space Force, like the Roman navies of the ancient world, will provide security and economic stability—of the space domain for the realm of space as the Romans did in the maritime arena (Shaw 2021). Technologically, one of the main drivers will be developments in rocketry (reusable), most likely combinations of propulsion systems from nuclear to magnetoplasma, and more—the first space mining technology, space-based solar, space-based communication, and beyond.

*Phase II: The Grand Strategy and Doctrine Era*

In the second period, the grand strategic period of 25 to 50 years, the goals are much more significant and have even more long-lasting and titanic consequences. If it can be condensed into one word, the watchword would be cislunar, and America's need to be more than first among equals in this arena. Space Force needs to be the stabilizer and protector of the new economic revolution and "liberal world order values" (Mozer 2021). Space Force will take on the navy's role in the nineteenth and twentieth centuries (C. Smith 2020). It was the 2040 period that the second Space Futures Workshop emphasized to help answer what the USSF will need in order to win. Multiple scenarios were discussed during the 2021 Space Futures for 2040 exercise: shaping, contesting, and fighting. As of this book writing, the official report has not been published. The reader should be reminded of these scenarios here as they illustrate the pitfalls and problems of the "Space Force after next."

1. **A Slip in Perseverance**—Loss of U.S. leadership in an expanding and increasingly valuable space domain. What happens when space is rich and hospitable, the United States is not the one running the place, and other space powers are achieving infrastructural breakthroughs?
2. **Claiming Paradise**—Sudden transformation in space value driven by black swan technological breakthrough. Extremely rapid propulsion and traditional propulsion and launch technologies coexist in a world where the space economy did not live up to its promises . . . yet.
3. **A Crowded Garden**—Conflicts in an expanding and highly multipolar space world. Multiple great powers within a flourishing space economy compete in the unforgiving terrain of gravity.
4. **Venus and Mars**—Piracy, revolution, and conflict in a unipolar space world. A blossoming space economy reveals the limit of defending an unconstrained spatial domain.

5. **The Expanse**—Cascading disasters strain actions and space power relations across an extended and expansive space domain.
6. **A Great Leap Backwards**—Failed international order in space in a deadly series of man- and nature-triggered space catastrophes impose colossal pressure on the space enterprise across the globe, uncovering buried tensions.

All of these fell under the concept of images of future operations (IOFOs). These IOFOs were constructed to span situations requiring U.S. action across the three phases of the conflict continuum: shaping the situation, contesting within the situation short of armed conflict, and engaging in the use of force to achieve U.S. interests.

These are the arenas that the Space Force will need to harness space power to be victorious. Space power strategy needs to be both general and specific.

On the one hand, it must be broad, coherent, consistent, comprehensive, descriptive, prescriptive, and combine the realism of today with the promises of the future (Ziarnick 2015). On the other hand, it needs to be specific toward tangible goals combining multiple missions, including securing the domain, extending from the Earth and beyond, and planetary defense. It will likely have a space-based weapons component that looks toward and beyond Earth (Carlson 2020, 198–199). It will need to have a complete picture of space domain awareness, reconnaissance, and surveillance. It will first need to dominate and govern the orbits of the Earth, cislunar space, and then the Moon itself. Finally, it will need to promote energy production such as space-based solar power, resource extraction, nuclear-powered spacecraft, and bases in and around the Lagrange points, the Moon, and beyond.

One of the significant defects among great power analysts and space power advocates is their inability to see the synthesis of the two. Many great power theorists see the world from a terrestrial level and focus heavily on the many national security threats that the United States faces. These range from terrorist organizations all the way to adversaries, nuclear arsenals, and everything in between. They will talk about the Russian/Ukrainian border or the Taiwan Strait. Equally, space power advocates will drone on about LEO and GEO and cislunar in a literal space vacuum from Earth-based problems. This compartmentalized thinking is a prescription for disaster. A genuinely effective space power doctrine and strategy that the USSF must follow will synthesize these. The DOD's 2020 *Space Strategy* guidance outlined a general strategy for this emphasizing four lines of effort (LOEs): (1) build a comprehensive military advantage in space; (2) integrate space into national, joint, and combined operations; (3) shape the strategic environment; and (4) cooperate with allies, partners, industry, and other U.S. Government departments and agencies (Department of Defense 2020a). Those of us involved with the 2060 Space Future strategy summed it up this way. "The U.S. must develop a long-term, national space strategy to ensure continued leadership. This strategy should be developed across government, industry, and academia to ensure synergy of efforts to optimize and promote overall U.S. national space power and grand strategy" (U.S. Space Command 2019). Space can no longer be treated as a subset for other national security and strategy aspects (Moltz 2019). It must be national security and strategy. In other words, space doctrine and strategy is national security writ large.

In simple language, space power is only necessary as it serves the United States and, next, allied interests. Space is not essential as an end in and of itself. First, this is a concept that many of my space colleagues need to accept. Just as the Louisiana Purchase was not necessary for and of itself, but as an enhancement of American security and prosperity, so it will be with cislunar space. Further, American interests are only vital as long as America stands as the bulwark of Western civilization.

Thus, the U.S. Space Force is merely an extension of the United States, and space strategy will rely mainly on the actions of the Space Force. If space strategy is the prioritization and amalgamation of military, legal, commercial, scientific, and civil activities, then it will be Space Force's job to be directly responsible for some of these activities such as military, and most likely legal, but to also create the environment for the others to flourish. Space Force will uniquely engage in warfighting, but no more uniquely than previous differences in warfare (Wirtz 2009). Space Force will need "full-spectrum dominance," not just preeminence. Just as some could not imagine steamships replacing sails or the development of the airplane, some cannot project this logic into space. Space Force will be the leader in ensuring the projection of American national power in total. The only other choice is for the United States to cede its primacy position and eventually become subservient to another great power that will take all of these advantages from space for itself. It is the fundamental job of any American military branch to dominate their domain of warfighting, just as it will be the job of USSF to dominate the space domain of warfighting and the clear recognition that space power can dominate the Earth in a way that earth power cannot dominate space (Carlson 2020). This was the fundamental trend from the Space Futures 2060 project (U.S. Space Command 2019).

Space power will enhance other military domains, as other military domains will enhance space, and both of them will assist in the NewSpace economy (Carlson 2020). This will begin with energy and resources and ultimately lead to exploring beyond our solar system. The command of space will assure national power. This will start with the management of access and communication and extend exponentially beyond dominance and governance. Initially, the power that controls low Earth orbit will be the power that dominates all terrestrial events and activity. This will be no different from terrestrial strategy, including gaining the high ground, the strategic and high-value positions, lines of communication and access, and the choke points. Thus, war and diplomacy in space will be consistent with humanity's historical experience in statecraft and power. This will be the realm of "astrostrategy" and the identification of "critical terrestrial outer space locations to exercise political and military dominance" (Dolman 2002). These are all more fluid in space but nonetheless real. There are immediate steps to take, including ensuring the American position in navigation, surveillance, reconnaissance, weather (including in space) forecasting, and missile warning. There are legal steps to prevent adversaries from using lawfare against us. There are deterrence steps and first offensive steps, including the deployment of space-based missile defense, ensuring American and allied dominance in LEO and then GEO, and the abolition of treaties that have violated the terms of their existence for decades, not least of which is the Outer Space Treaty.

One thing is abundantly clear: the only path for serious national security for the United States is to be the dominant space power.

# CHAPTER 4

# Space Intelligence

This book is highly conceptional and perhaps speculative. The discussions about future wars and conflicts have yet to occur. Any existing technology that could define such conflicts remains at the embryonic stage or is only notional. In many ways, this chapter on space and intelligence is the ultimate example of all of this. Critical to this discussion is that we must shift from seeing intelligence from space to intelligence *for* space (Raymond 2021).

In 1963, Allen Dulles, director of the CIA, defined intelligence as "Foreknowledge, a kind of prophecy—like craft, which is always on alert, in every part of the world, toward friend and foe alike." Intelligence is three-pronged. It is a process (means of collection, analysis, dissemination) that creates a product (analysis, reports, briefings that are "actionable") under an organization (collection of units or agencies that carry out intelligence work). Most intelligence agencies have an analytical wing and/or a collection wing. Many have a covert operation arm that ranges from mild collection efforts by diplomatically covered officers all the way up the spectrum to paramilitary operations often linked with military special operations. The most important roles of intelligence are strategic foreign intelligence, counterintelligence, and covert action. It goes without saying that intelligence, as former DCI Jim Woolsey explains, is of "great importance" (Woolsey 2021).

## SPACE INTELLIGENCE IN THE PAST

The last major reorganization of our nation's intelligence community (IC) was in 2004 when the Intelligence Reform and Terrorism Prevention Act removed the director of central intelligence (DCI) from the top role of the IC and replaced him with the director of national intelligence (DNI). Although the IC has 18 robust intelligence agencies, 56 government bureaus exist that perform some aspect of intelligence. Space has been a mere subset of intelligence. The real action was terrestrial, and pop culture consistently focuses on the covert activities of the CIA. At best, space was considered a supporting actor to the real business of intelligence. The grand question regarding intelligence and space, particularly Space Force, is what will space intelligence look like? In January 2021, USSF became the 18th member of the American intelligence community (U.S. Department of Defense 2021).

However, when most national security experts think of space and intelligence, they almost entirely only think of the National Reconnaissance Office (NRO) established in 1961, which openly states their mission: "The NRO is the U.S. Government agency in charge of designing, building, launching, and maintaining America's intelligence satellites" (National WW2 Museum n.d.). In partnership with the National Geospatial-Intelligence Agency, they give the United States a space-eye view of all threats. Another legacy organization from the 1960s Cold War is the Missile and Space Intelligence Agency (MISC), whose primary task is the technical side of weapons platforms, particularly missiles, and is under the DIA.

In 2022, Space Force activated the National Space Intelligence Center, whose stated job is "NSIC will perform national and military space foundational missions and will evaluate capabilities, performance, limitations, and vulnerabilities of space and counter-space systems and services" (Underwood 2021).

## MANY VIEWS, FEW SOLUTIONS

It is clear from the multitude of sources and interviews for this book that the national security and space community is divided over the future of intelligence and space. In 2009, David Arnold summed up the status quo thinking on the topic when he wrote, "Space-based intelligence is a field that focuses, during peacetime and hostilities, on the accumulation, analysis, and dissemination of information about the enemy terrain, or the weather in an area of interest" (Arnold 2009).

This is the status quo thinking. The job of space intelligence is to provide the IC with data for terrestrial-based problems and conflicts. This, to be sure, is a critical component of intelligence. However, it is green water thinking in an era described in chapter 1 that will be determined by black space thinking. In our 2040 space strategy workshop, the gaps in intelligence need were powerfully illustrated as scenarios played out against great power adversaries, foreign terrorist organizations, criminals, and mid-power coalitions. The need for real-time space situational awareness, space domain awareness, space human intelligence, and covert action proved the problems with a Balkanized approach to space intelligence. It proved that a government primarily led by Earth-centric leaders would doom the United States to being in the dark figuratively and literally when it comes to space. The current "satellite-centric thinking" must transition to a new mindset, risk culture, and organization (Carlson 2020, 88). The area of space and intelligence will be one of those clear areas where Space Force will need to overcome the status quo thinking of the military hierarchy (Ziarnick 2015, 63). Put simply, space intelligence needs its own professionals as the other services and branches (Shaw 2021).

As with everything in this book, the first steps will be in LEO, where the United States must first establish a complete situational awareness as quickly as possible (Klein 2006, 221). However, as necessary as this is, it is only the nascent step to space intelligence. Thus, we are at a significant dichotomy. The concept that we need to dominate LEO and the Earth with surveillance is but the beginning of a vast kingdom of intelligence that needs to be created. Still, many national security and space communities believe it is an endpoint. General Steven Kwast put it best: "But ultimately, we as a nation, Congress needs to re-look at the design of the intelligence community and make sure we have the right tools to understand what is going on around us in cislunar space. Right now, we can't see what is going on, on the far side of the Moon . . . we are blind. . . ." We must ensure "that the intelligence community has a strategic design that folds space and cislunar space economy into its structure" (Kwast 2021).

## THE INTELLIGENCE WE NEED—A SERVICE FOR THE STARS

I have heard and read the myriad of thoughts about the issue of space intelligence. I fully understand the arguments about not adding another bureaucracy to an overly redundant and bloated intelligence community. I also realize that we already possess the NRO, NGA, and elements inside the IC devoted to space analysis. The USSF is launching a new national space intelligence center. It is clear that many space power professionals do not estimate that the current IC can fully handle the new frontier regarding space and see the NRO's existing housing as a problem and that "America needs dedicated career space intelligence professionals to support civil and military space operations" (C. Smith 2020). The IC has space-centric professionals, individually. However, none of this will handle the national security needs in the area of intelligence that the United States needs. It will not be an easy task to reorganize the IC to do this (Carlson 2020).

The United States must do one of two things. It either needs to exponentially expand the role of the USSF intelligence branch or, if that is out of the question, create a U.S. Space Intelligence Service (SIS) that would be an independent intelligence organization entirely devoted to anything related to intelligence from LEO to the stars. It might be possible to create this within the CIA. However, it would most likely diminish the CIA's ability to complete its core, terrestrial-based mission and only give space intelligence a portion of what it needs. This expanded Space Force or SIS would need to absorb the NRO and NGA, and they would need to become an "inward/outward"-looking organization. This intelligence service would need to be divided into a leadership directorate, ranking at the same level as the DCI, an analytical division, a collection division, and a covert division that would need to be linked to a USSF special operations division. Some suggest that USSF should not have a special operations branch. However, every branch of the U.S. military, including the USAF, has special operators that engage in specialized and or covert operations. It defies all logic that Space Force would not have the same. It would also be critical to have space-centric counterintelligence professionals who will need to deal with the myriad of threats from rival intelligence services that will also be operating in the space environment.

Space professionals are the only ones who will be able to understand the unique conditions for space that will change how we collect intelligence, analyze it, and take action in it. But unfortunately, it will take years for an intelligence analyst to understand this complexity, and to be distracted by other assignments inside a terrestrial-based community will mean that we will always be behind our adversaries.

In our 2040 scenarios, severe, almost catastrophic deficiencies were discovered related to intelligence. These deficiencies centered around domain awareness, surveillance, the ability to assign attribution, infiltrate threats, understand motivations, operate in the space environment, engage in espionage and paramilitary activity in the space environment are going to require training and skills that almost no one in the current IC has or will likely have in the near future. The private sector is keenly aware of the problem of protecting space-based intellectual property as creator, inventor, and innovator in NewSpace. Ian Molony, a pioneer in areas like space architecture, design, and invention, states, "It will be paramount in assuring that long-term assets, manned or autonomous scientific or commercial outposts on the lunar surface, satellites or orbiting utility, communications, and scientific platforms, remain secure from espionage, physical attack, space piracy, and computer warfare" (Molony 2021).

All the arguments made as to why the U.S. Space Force needed to be separate from the U.S. Air Force are exponentially genuine in the realm of space intelligence.

Related to this intelligence role, Space Force will also need to play a constabulary role merged with its intelligence wing. If the United States does not have a space law enforcement arm, we will see the end of the space economic revolution before it begins. As General Steve Kwast has stated, this will be the only way to "ensure lawful and non-hostile users of space enjoy the same freedom of navigation that our navy ensures on the high seas" (Carlson 2020, 4). This is exactly the role of the U.S. Army during western expansion, particularly the U.S. Cavalry. There was no FBI or coherent local policing system. There won't be in space for a very long time. The military, throughout history, has had to play the role of the "watch," the "constabulary" when none existed. Our notions of law enforcement need to be redirected to the past to accomplish the future.

There is much talk about how space intelligence is necessary to protect the United States from a space Pearl Harbor. However, there is an equal need to protect the United States from a space 9/11. The United States won't just face national threats in space, and the space economy will also fall victim to terrorists, criminals, and pirates.

Let the reader not forget that until 1947 there was no agreement on the need for a permanent intelligence service. In fact, it was fought against vigorously. Some of these arguments were very valid and based on a republican view of liberty that opposed a permanent and professional military and intelligence system. We had an ad hoc intelligence service during the Revolution, the Committee on Secret Correspondence. This was disbanded until the Civil War when a combination of the Bureau of Military Intelligence and the Pinkerton Company provided an assortment of intelligence. This was also dissolved. The uniformed intelligence services that rose up provided some narrow service-related intelligence, and the Department of State had some functions. During World War II, the Office of Strategic Services (OSS) was created and then disbanded at the end of the war.

The United States lacked a consistent and professional service until it was thrust upon us with the realization that enemies like the USSR would not simply vanish. It was not until the early twenty-first century that the IC as a whole took the Islamic terrorist threat seriously, regardless of the countless warnings for decades. Let us not pretend that the IC can make good national security decisions by itself. This backward-looking mentality was part and parcel of some of the worst disasters in American history, ranging from the assassination of President Lincoln to the attacks on the Twin Towers and beyond.

Finally, no space professional believes that American interests will stop at LEO or GEO. It is already accepted that American interests extend to cislunar space and the Moon. Elon Musk is already planning to colonize Mars. It will only be a matter of time that American interests extend throughout the entire solar system and beyond. The need for a space intelligence service either as a significant component of USSF or as a separate service is a logical certainty. The only question remains will we allow a Space Pearl Harbor to force us into it.

# CHAPTER 5

# Allies, Diplomacy, and International Law

Critical to America's push into space, and therefore of Space Force's role, will be maintaining and building coalitions and alliances that protect the vital and national interests of the Western powers and the values of Western civilization and democratic ideals. This is sometimes referred to as the international liberal order in international relations, but a better explanation is the Pax Americana.

America has led the West since 1945 as one of two superpowers until the fall of the evil empire in 1991. From 1991 onward, the Pax Americana has enforced an empire of liberty that has created a multitude of alliance systems and coalitions. Examples of this would be NATO, ANZUS, AUKSA, the defense pacts with Japan and South Korea, and the unique diplomatic situation with nations like Israel and Taiwan. The core of this is naturally the Anglosphere, which is often referred to as the Five Eyes in realistic terms.

It will be necessary for the United States to take this democratic alliance into space under American leadership. Therefore, it is essential for the United States in general and the Space Force to lay the foundations for hard-power and soft-power partnerships to ensure that space remains a zone of liberty under law, free-market economics, and open access. But, unfortunately, American leadership is a double-edged sword requiring massive obligations on the part of the United States, specifically the Space Force, and reciprocal commitments from our allies.

## GLOBAL NORMS IN SPACE

It is incumbent on the United States to be first in setting agreed-upon norms internationally with the ability to enforce them. As previously discussed, all spacefaring nations have militarized space already (Dolman

2002, 2). Many believe that assuring space access will lead to peace, assuming they believe in the peaceful use of space (Gallagher 2010).

There are multiple templates for creating norms in space. One of those templates would use international law to create space law. First, however, there needs to be an understanding of when terrestrial law ends and begins again. In other words, the area between American law over the sovereign territory of the United States and American sovereign territory in space and on regions in space needs to be governed by a form of international law. There is much debate about where this starts and ends. Many wish to use the Karman line at about 62 miles above the Earth's surface. Suppose there is going to be international space law. In that case, it will need to accept many of the terms of international law such as the role of innocent passage, the sanctity of medical ventures, and the treatment of prisoners of war.

Further, we will have the problem of sovereignty. Perhaps the only tentative and working definition at this point is a place that can actually be defended (Dolman 2002, 118).

"International law" is a term misunderstood by policymakers, academics, and the general public. Much of international law, hence future space law, is based on diplomatic custom and culture, which are rarely enforceable on belligerent lawbreakers. Even those parts governed by conventions and treaties can only be enforced by a signatory willing to use force or coercion.

Treaties are the "hardest" form of such law, the nature of which was set out in the 1969 Vienna Convention on the Law of Treaties. They are binding and based on the good faith of the participants. Custom is somewhat softer but includes agreed-upon norms such as diplomatic immunity and the sovereignty of foreign embassies.

In general, conflict over future space norms will probably focus on the same Earth-based international law sources, namely, jurisdiction and sovereignty. In general, outer space is thought of as the high seas where claims of sovereignty, denial of access, and restrictions on use are prohibited.

The single most significant debate on overall space norms will come from the core debate about the law in general. Western civilization—in particular, American civilization—has endorsed an international law based on the "naturalist" school, arguing that there is a universal moral order under the fatherhood of God, and individuals, groups, and nations are subject to it. This is the basis of the war crime trials following the Second World War, as an example. In contrast is the positivist school, which argues that law is human-made, enforced only by human authority, based on the consent of each nation. This is primarily where America's adversaries, such as Russia and China, take their stand. This natural law/positivist debate will likely be the bedrock problem for space governance in the future.

International space law as it is currently known can best be described as three treaties dating from the Cold War under the UN umbrella and a fourth attempt to demilitarize space and create a broad international regulatory framework. This last one is not ratified by any significant world power. The UN Office manages the three major UN conventions for Outer Space Affairs. The Artemis Accords, which were signed first signed by the United States and several partner nations on October 13, 2020, appear to be a push by the United States and allies to set its interpretation of UNs space treaties.

The first of these UN treaties is the Outer Space Treaty (OST) of 1967, preventing the placement of WMDs in space and forbidding national claims on space territories. The 1967 OST is considered gospel by many involved in space who refuse to recognize the treaty for what it was (an attempt by the United States to control the situation at the time) and as a relic of an age that no longer exists. It was inherently flawed from the beginning. Like many treaties, it must be governed on one of the foundational principles of international law dating from the Greco-Roman era: *Clausula rebus sic stantibus*. This concept means that a treaty is null and void when the terms of said treaty no longer exist. Another legacy of international law is *Cuius est solum, eius est usque ad coelum et ad inferos*, meaning whoever's is the soil, it is theirs all the way to heaven and all the way to hell. This is essentially the realist argument for establishing sovereignty in space. Other recognized norms would be a sustained occupation, charting and mapping, naming locations first, and the priority of discovery or, as some call it, first presence. Terrestrially, these disputes are numerous: the Korean peninsula, Antarctica, Israel, the Senkaku Islands, the Falklands, and many others. Space will be no different.

Let us be clear to the reader. Treaties are not holy writ. They are human-made creations, more often than not based on realist hard-power assumptions by powers that have an interest, at that particular moment, in signing them and *hopefully* following them. Thus, America should follow the mechanism for withdrawal from the Outer Space Treaty just as President Bush withdrew from the ABM Treaty and President Trump withdrew from the INF Treaty. These are treaties that our adversaries did not follow, and only resulted in reducing national security for the United States.

Three other essential space treaties are the Agreement on the Rescue of Astronauts, the Return of Astronauts, and the Return of Objects Launched into Outer Space of 1968; the Convention on International Liability for Damage Caused by Space Objects of 1972; and the Convention on Registration of Objects Launched into Outer Space of 1976. These treaties govern space, making the launching nation responsible for their space objects.

Finally, there is the rescue agreement defining parties that are obligated to render assistance. The Agreement Governing the Activities of States on the Moon and Other Celestial Bodies, the so-called Moon Treaty of 1979, goes unratified by any major spacefaring nation like the United States and China. The Moon treaty would effectively ban resource harvesting, colonization, and claims on the Moon. The Prevention of an Arms Race in Outer Space of 1981 was also unratified and opposed by the United States. This would attempt to ban all weapons in space, not just WMD.

In contrast, the 2020 Artemis Accords is a serious cornerstone for space norms. It is an agreement between countries, including participating in the Artemis Lunar Program. Initially signed by senior officials for space policy for each nation—Australia, Brazil, Canada, Italy, Japan, ROK, Luxembourg, New Zealand, Ukraine, UAE, UK, USA, Isle of Man—it is a multinational agreement centered on the Moon but with implications for other celestial bodies. It includes an agreement that extraction and utilization of space resources should be conducted in a manner that complies with the Outer Space Treaty and supports safe and sustainable activities. The Artemis Accords will also be an avenue for alliance members to partner with Space Force (Goswami 2021). The signatories affirm that this does not inherently constitute national appropriation, which the Outer Space Treaty prohibits. They also express intent to contribute to multilateral efforts further to develop international practices and rules on this subject (United Nations 2020).

The accords are widely viewed as an attempt for the United States and its allies to set its definition for mining the Moon and other bodies (Newman 2020).

On the other hand, our adversaries are attempting to create positivist norms for their own benefit.

Allowing China to continue on a path to prominence in space is dangerous given that nation's propensity to ignore international law. For example, the Moon offers valuable commodities at specific locations like water in the polar regions. According to a report by the U.S. China Security and Economic Review Commission, "in 2015 Ye Pejian, the head of China's Lunar exploration program likened the Moon and Mars to the Senkaku Islands and the Spratly Islands, respectively, and warned not exploring them may result in the usurpation of China's space rights and interest by others," which demonstrates that Beijing is already focused on claiming Lunar resources (Davis 2018).

The Moon may be the place where alliances, coalitions, and treaties first come to a head. The great powers are all interested in gaining an economic and strategic foothold on the Moon, establishing bases and ultimately colonies. For example, China has declared its intent to have a permanent moon base by 2024 and expand it through 2027. China is obsessed with the notion of the first presence. This would allow it to pick the best locations and use a loophole in the Outer Space Treaty allowing nations to create structures and equipment, thereby blocking others from doing so, yet not actually claiming sovereignty (Mishima-Baker 2021).

## ALLIANCES AND SPACE FORCE

The rise of potential adversaries, foreign terrorist organizations, criminal cartels, and other bad actors will be an obstacle to the United States in space, and alliances are a large part of the solution. These alliance systems add "depth" to American national security (Gingrich 2021). General Raymond laid out the broad goals: "First of all, our goal is to deter conflict from the beginning or expanding into space. And I think alliances help us with that. I think the other thing is that space is a global domain and requires global presence. The other thing that we want to do with partners is developing standards for safe and professional behavior in the domain. And we're working that very closely with them. The design of our force structure should be partner-friendly from the beginning, and we can have more than just countries help us pay for some capabilities. Ideally an integrated force design would make it much harder and complicated for targeting by a potential adversary" (Raymond 2021).

The space environment will be no different, with the freedom-loving world looking to the United States not only to protect them but to create an astropolitical system of stability and consistency. The United States must "establish the conditions under which these rising space nations will want to partner with the U.S." (Mozer 2021). Like any diplomatic alliance, it will come with cost and benefit and will need to be balanced against specific American national and vital interests. As Coyote Smith says, "First, international partnerships are used as a means of expanding the bureaucracy surrounding acquisition programs to make them harder for politicians to cancel. Second, an ally will contribute something substantial to an American effort from time to time. This is rare and usually comes as a result of a diplomatic trade-off. Finally, international partnerships contribute to a sense of legitimacy, credibility, and acceptability of a program or activity among the broader international community" (C. Smith 2020). General Steve Kwast laid out the solid case for ensuring that American interests drive the mission rather than letting the mission drive the interests. "If America has to do something to protect our survival or our values, whether it's our economy, our government or our people, our sovereign soil, America has to have the unilateral capability to be able to do it alone if nobody is willing to join. So as we invite the rest of the world to join us on this journey and everybody can contribute to whatever degree they can afford to contribute, both in resources and in will and political willpower, we as an American society need to make sure we do not outsource the key capability that we rely on to another nation. Because history is full of examples where our friend today becomes our adversary tomorrow" (Kwast 2021).

Some of the challenges with military alliance partners today are interoperability, consistent levels of training and technology, and leadership styles. These will all have to be addressed anew in space (Carlson 2021). However, aspects of these are already being addressed, such as with the Combined Force Space Component, with America's Five Eyes partners, and additionally, Germany, France, and Japan modeling on the combined air and maritime operations (Shaw 2021). The recently published *Combined Space Operations Vision 2031* attempts to create an alliance-based mission in its primary statement: "Generate and improve cooperation, coordination, and interoperability opportunities to sustain freedom of action in space, optimize resources, enhance mission assurance and resilience, and prevent conflict" (Department of Defense 2022). It sets out a brief conceptual roadmap for the Western alliance to share common principles such as freedom of use of space, "partnering while upholding sovereignty," unity, and the sharing of intelligence.

To reduce chaos and increase effectiveness, it will be necessary to create a single space-based alliance system that knits together the democratic multilateral and bilateral arrangements between the United States and its Western allies. The most successful international order maker has consistently been the United States and the Pax Americana. However, strategic decisions must be made to ensure the continuance of global and astropolitical stability. One of these decisions concerns the future of American international leadership: a dynamic international system for the NewSpace era must rise to meet such challenges. This is a system where the United States spearheaded the creation of amplification of NATO by fostering the DAN—Democratic Alliance of Nations.

A new organization with credibility with the American people could go a long way toward solving all of these issues. This new dynamic organization could go by many monikers; the one used here is the DAN. The DAN would model itself on NATO, and if successful, could replace NATO and stop the endless bickering about the future of that historically critical organization. The fundamental essence of NATO would remain; this would include a supreme commander who would be an American, a rotating political head, and Article 5 would serve as a similar trigger for action. However, there would be some drastic differences as well. Only nations that were willing to employ proportional military force (not token support) would be allowed membership. The Article 5–style trigger of "an attack on one is an attack on all" would be broadened. These triggers would have to include preemptive and preventative threats, as well as a mechanism to deal with genocide, massive human rights abuses, regional despotism, rogue states, and failed states.

Further, there would have to be a clear mandate that military force would be used under these trigger conditions. This does not mean that military force would be the first or only option, but it would be a viable coalition response, unlike the Security Council. Critical to the composition of the DAN would be that membership is reserved solely for true democracies. This would be subject to the scrutiny of the founding members and include such benchmarks as working democratic constitutions, the actual rule of law; a vibrant civil society; the complete protection of private property; and obedience to natural law. The organization's foundation would start in the Anglosphere (U.S./UK/Canada/Australia/New Zealand), NATO, and ANZUS. Membership would hopefully be expanded to states such as (but not limited to) those in Western and Central Europe, Israel, South Korea, Taiwan, and Japan. It goes without saying that some of these nations would need to make fundamental changes in their foreign policy, legal mechanisms, and even political culture.

Therefore, it would be through the DAN that the United States could lead the free world in a dynamic twenty-first century, as it had through the tribulations of the twentieth. Second, it would further the security of the United States and the American people. Third, it would illustrate to the electorate that the United States is not forced to act alone or bear the only burden. Finally, it would further enhance members' political and economic connections for stronger ties and bonds. Finally, it would be the stepping-stone for Western democratic values to the stars.

This type of alliance will eventually need to take on a name, like NATO. However, this Democratic Alliance of Nations will serve as our placeholder for now.

It is clear that just as in the geopolitical realm, American primacy and leadership will be required to achieve a positive future not just for America but for all humankind.

## CHAPTER 6

# The Triplanetary Economy

The creation of the U.S. Space Force in late 2019 as a free-standing branch of the American armed forces and concurrently the U.S. Space Command's reestablishment has brought new and much-needed attention to a critical emerging domain: space, specifically space economics in the coming economic revolution. The 2020 DOD report recognizes five fundamental challenges that space presents to the military: new problems of power projection, a lack of space experience, the ability to work with international partners, the ramping up of activities by our adversaries, and a general lack of public understanding (Department of Defense 2020a). This could not happen at a worse time for the new economy on the horizon. We need to disabuse ourselves of thinking that the role of the military and the new economy are separate. They are the same for those wishing to protect and enhance America's vital and national interests. We can easily see the roots in expeditions like Lewis and Clark, primarily a military mission. Equally, the USSF will continue this military support for commercial and scientific exploration, similar to our activity in Antarctica (Carter 2021). The watchword that continues to come up repeatedly among senior officials regarding the USSF's economic role is stability, with the template being the analogy to the U.S. Navy (Raymond 2021).

A new economic strategy and new economic thinking begin with mapping cislunar space, a domain that encompasses a near-space environment between Earth and the Moon. Cislunar space is the space between the Earth's atmosphere and the area right beyond the orbit of the Moon. Strategically, cislunar includes the Lagrange points, the points in space where there is an equilibrium between Earth's and Luna's gravitational force. It is an area that holds military, political, cultural, and economic consequences that will determine the success or failure of American strategic primacy for the twenty-first century and beyond. Any discussion of space development needs to begin with a fundamental question: Is it worth it?

The United States is late to recognize space's commercial and economic potential. Much of this has been due to the lack of political will and poor decision-making (Dolman 2002, 136–137). While the projected evolution of a space economy is still conceptual in nature, it is already clear that the benefits of space development will be innumerable. As it stands, the realm of space represents a nascent—and as yet mostly untapped—market. An increase in space satellites would facilitate a faster and more reliable internet on Earth and reinforce the speed and reliability of calling and messaging on terrestrial telecom networks. Shipping capabilities in space could also be a growth industry, and eventually, blasting a payload into space will become less costly and faster than shipping across an ocean or via commercial aircraft. Asteroid mining is also a likely space industry, and potentially an extremely lucrative one.

## RESOURCES

Resources often scarce on Earth are many times more plentiful on asteroids across the galaxy. The website Asterank estimates over 830 asteroids worth over $1 trillion with minerals such as cobalt, nickel, and platinum (Asterank n.d.). The 2015 U.S. Commercial Space Launch Competitiveness Act goes a long way to further American commercial interests. "A U.S. citizen engaged in commercial recovery of an asteroid resource or a space resource shall be entitled to any asteroid resource or space resource obtained, including to possess, own, transport, use, and sell it according to applicable law, including U.S. international obligations" (114th Congress 2015).

The limitlessness of these resources is essentially priceless; "like Mackinder's heartland, [it] is so vast that, should any state gain effective control of it, that state could dictate the political, military, and economic fates of all terrestrial governments" (Dolman 2002, 68). However, these resources' boons will inevitably lead to conflict over who can exploit them. The idea that America can allow "nature to take its course" flies in the face of human history. The American military will be required to help secure and protect these resources or prevent them from falling into the possession of malevolent actors. Although the romantic concept of space as a collective commons is dangerous from a security perspective, it will also severely stunt the American private sector's ability and willingness to do what is economically necessary.

Over time, space transport will revolutionize the global economy and several industries in particular: aerospace and defense, IT hardware, telecom sectors, space tourism, package delivery, and energy. In addition, we will witness an interplanetary superhighway from Earth to LEO, a lunar transfer orbit (LTO) to the Moon, low Mars orbit, and then Mars (U.S. Space Command 2019). However, it will be energy that might have the most immediate and direct impact on the lives of Americans, as it will eliminate many of the problems surrounding climate change and conflict over fossil fuels. This will be especially true if America becomes the leader of space-based solar power, with "a total of 173,000 terawatts (trillions of watts) of solar energy [striking]s the Earth continuously" (Chandler 2011).

Creating strategic trade routes in "near space" will hearken back to the change in trade and globalization during the Renaissance. The term "globalization" will need to be altered to include space, and a new term such as "cosmosization" or "interplanetary commerce" will replace it in more than name. The NASA Gateway project, which will be built in orbit around the Moon, will deliver goods, services, and personnel to and from the lunar surface (NASA 2018b). This mission depends on the Space Launch System (SLS) and Orion spacecraft.

## THE NEW ECONOMIC REVOLUTION

This new economic revolution will grow exponentially from its inception. From the private-sector side, it will be dominated by NewSpace companies, defined by visionary Rick Tumlinson as having the "traits of Silicon Valley" start-ups. They are lean, mean, and fast-moving. They are also not usually designed primarily for government contractors—although they often do government contract work or accept government contracts. Thus they are more structured as "commercial" rather than "cost-plus entities" (Tumlinson 2020). Morgan Stanley estimates that the "space industry" will generate $350 billion annually, a figure that could grow to $3 trillion a year if this system begins to be implemented (Morgan Stanley 2020). Predictions of exponential factors of GDP and multitrillion-dollar markets are more than possible (Goswami and Garretson 2020, 8). This wealth will benefit the United States and humanity as a whole (Dolman 2002, 6). This, however, does not even account for the upward changes created by the technology developed by the space economy as it evolves. In order for this to happen, an industrial reorientation is necessary. Simply focusing on exploration and scientific discovery are not sustainable economic and strategic models purely in themselves. Development of the near-space economy will require economic and industrial output and innovation that will fundamentally change the international economic system in ways not seen since the Industrial Revolution's transformation. It will also need mankind to reorient its economic system as a whole. This imperative may be complex for many people to grasp, but it is also why America will have the best chance to lead this new economic revolution. Eventually, this new economic revolution will lead to colonization and settlement. We already have American models from the High Frontier envisioned by Gerry O'Neill (O'Neill 2019). and even the nineteenth-century models of manifest destiny (Impey 2016).

## AMERICAN SECURITY AND THE NEW ECONOMY

America landed men on the Moon and answered the call of President Kennedy's dream. American culture and history are infused with such concepts as pivot, adapt, and innovate. American culture will need a rebirth of the frontier spirit and a declaration of what famed historian Frederick Jackson Turner said in 1893, that the closed frontier had been reopened on a much grander scale. This, in turn, requires us to tackle security in space seriously. The new commercialization of space and the ultimate low cost of launches and satellite proliferation create new security challenges (Department of Defense 2020a). It is estimated that by 2028 there will be 15,000 satellites in orbit (Wood 2020). We will also need to overcome long-term scientific and technical challenges, such as new technology for space superiority, the resilience of space systems, access and logistics, and the integration of a myriad of disciplines and domains (U.S. Space Command 2019). Many of those who oppose the Space Force did so because they see space as primarily a realm of exploration and scientific interest. Yet if we want it to become more, such as an arena for commerce and innovation, we must ensure its safety and security. That, in turn, requires a new type of military thinking. However, in some ways, it hearkens back to yesteryear, where Space Force will often play the role of the nineteenth-century U.S. Cavalry on the frontier as the only natural source of law and order (Gingrich 2021). In that vein, it must look at various initial missions that assist the new economic revolution. "The Space Force must serve as both shield and pathfinder. As shield, the Space Force must protect friendly, especially industrial and commercial, entities from hostile or pirate forces. Without a guarantee of safety, commercial entities will not operate in an area. It must also shield against environmental dangers and offer life-saving and other emergency services to craft operating in cislunar and beyond. The Space Force must also serve as a pathfinder, which is to say it must break ground for the technology required for operations in space and share it with friendly commercial and military forces, lowering their entry costs to assist in accelerating the development of the new space economy" (Carlson 2021).

It is a great benefit for the private sector to develop the space economy and the needed technological tools. The military can modify these commercial products for their use. Again, we return to the idea of the virtuous cycle. Those that are serious about space power are serious about the new economy. It may require us to return to some of our economic thinking of the Founding Fathers' period that understood the national security needs to enhance and protect certain markets and technologies. Many advocate that this "enlightened industrial policy" is nothing new but will likely be necessary (Ziarnick 2021). In the context of space, security can be viewed on two levels. The first is international security—the security of the international system as a whole. The second is the security of the Western alliance: Western nation-states (including the United States), their allies, their economy, their values, and their political culture. If carried out by America and its partners, a serious plan for the former will necessarily serve to bolster the latter. Currently, the global system has no protection against an extinction-level event, nor is there an alternative for human civilization to escape a disaster. Space Force will also need to know which particular economic developments may pose a threat to American national interests and act accordingly (Garretson 2021). At first blush, this state of affairs may seem acceptable. Still, it becomes decidedly less so once one grasps the dangers posed by asteroid collisions, a Carrington Event (solar storm), and several other existential dangers. The current coronavirus pandemic provides a case in point; although far from a civilization-ending event, the disease has nonetheless illustrated the weaknesses, vulnerabilities, and gaps in our ability to protect national populations, as well as the fact that there is no alternative but to do so. The same holds for space. As the United States moves more and more into the space domain, the imperative will grow for the developing economy to be protected. On a mundane level, it will need to be protected from space debris, which can wreak havoc on space-based technologies such as satellites. There will also need to be a defense against a breakdown in communication or travel. But other security needs must prevail as well. No economic system can viably exist without adequate safeguards. That is the reason nations, irrespective of political and ideological outlook, have uniformly created penalties for threats to private property, penalized breach of contract, and provided security from hostility, violence, chaos, and criminality. There should be no doubt that a new economic revolution in space will foster the same challenges. From the potential of electronic disruption to the (currently fanciful) notion of space piracy, the space domain will assuredly face potential criminality and sabotage as it develops. Assuring that this disorder stays at a minimum will go a long way toward instilling confidence in the emerging space economy. This is already a great concern to the Space Force and needs to set behavior and norms. As General Shaw states, "I would start first and foremost with economic and commercial activity. Right now, we really don't have well-defined norms of behavior in space equivalent of, 'right-of-way' space. We have that all over the place in the maritime domain and the air domain. We wouldn't have safe operations for commercial activity in those domains without some sort of understanding of what's appropriate behavior and what isn't, what's the right way to enter and leave port or the right approach to a controlled airport. We need to do a better job of that. That will enable those actors that are operating in space to feel more secure and safe, to incentivize further investment and make operations a bit more transparent and predictable, just like we do in the other domains. That will also feed into national security activities, to some extent. It will be understood what normal behavior is and then in times of national security interest when abnormal behavior may become necessary. And I think we need to do that relatively soon. A second goal we need is called space domain awareness. It is just a better understanding of what's going on in the space domain" (Shaw 2021). There will be a synergistic cycle of space commerce affecting space activities, affecting space strategy with both merchants and Guardians mutually supporting each other to stabilize this new economic system (Klein 2006). The USSF should overtly protect the commerce of space as a fundamental mission (Mozer 2021). This may be easier said than done. Some public- and private-sector personalities and institutions have created a wall of illusion between economics and national security, which was never a reality from the founding of the Republic until today. "There are many out there that philosophically are trapped in that mindset, where you don't want the military linked with anything economic. But they have forgotten about what the Navy does in the open ocean, and in the commerce of the open ocean. The Navy of cislunar space will bring anybody to justice that violates, whether it's American sovereign law or the law of open oceans" (Kwast 2021).

Therein lies the conceptual case for a more robust American military presence in space. On June 18, 2018, President Trump changed the space dynamic by ordering the DOD to create a new sixth branch of the military entitled the Space Force, whose job will be to unify American national security concerns regarding space. It was created as part of the 2020 National Defense Authorization Act. The current mission is to train, equip, and organize forces for space. In August 2019, the United States reactivated U.S. Space Command as a unified combatant command whose job currently is to "conduct operations in, from, and to space to deter conflict, and if necessary, defeat aggression, deliver space combat power for the Joint/Combined force, and defend U.S. vital interests with allies and partners" (Morehouse 2022).

**THE NEW TRIPLANETARY PROJECT**

The Triplanetary project, encompassing the Earth/Moon/Mars, is an idea that recognizes that the strategic future of the United States in specific, and the Western alliance in general, is not confined to cislunar space, and extends out to Mars as a way of ensuring prosperity for humanity. The name is more for literary purposes rather than literal, as the Moon isn't a planet. Still, the project itself envisions a future of robust commerce and safe human transport spanning the range of space between Earth and Mars. Space explorers, colonizers, and entrepreneurs see Mars as the future crown jewel. "NewSpace" advocates view Mars as the initial epicenter of a serious human presence among the stars. However, several developments need to be completed for this dream to become a reality.

The first stage of a Triplanetary economy would be an exchange of goods and services between two Earth-based entities in space (the Earth and Moon). For example, an asteroid mining company may lead the economic impetus to send raw extractions to a "floating" base or to a Moon-based processing plant where the minerals and metals can be extracted and used. Future stages would expand from low Earth orbit (LEO), geosynchronous equatorial orbit (GEO), and cislunar space onto the Moon and then Mars. Basic resource use will eventually become trade, communication, energy production, and finally move from a human presence to colonization. Ultimately, this will set the conditions for an even farther interstellar exploration and expansion stage.

The Moon is a stepping-stone to the future, but Mars will be an important next objective as it has comparatively more to offer for human colonization. Solar energy can generate power on the Moon and Mars. Still, Mars has the possibility of wind power and can support agriculture and create a more "indigenous" civilization than can the Moon. Mars has the potential for rich and profitable mineral supplies, especially deuterium—a fundamental element for nuclear power, particularly with the promise of fusion. Furthermore, as this is being written, there is ample discussion about "terraforming" the environment to eventually make it possible to create a stable civilization on a place like Mars, which scientists think could be rich in nitrogen, hydrogen, carbon, copper, sulfur, water, and ice. The NASA Gateway project and the Trump Moon-Mars Development project provide early glimpses into the possibilities that an economic zone that encompasses Mars has to offer (Gingrich 2019). The Triplanetary project will be the launchpad to a permanent human presence in this arena and beyond. Throughout their existence, nations encounter pivot points—moments where they can choose between disaster and surrender or triumph and victory. A failure to expend the needed time and resources to plan for the future can lead to military disasters and civilizational downfall. History is rife with such examples: Athens during the Peloponnesian Wars (404 BCE), the Roman Empire in 476 CE, nineteenth-century China (which suffered three stunning military defeats, in 1842, 1860, and 1895), France in 1940, and so on. In each case, there was a failure to appreciate the technological and strategic advancements that no longer conformed to past doctrine. History is littered with those who lack the requisite foresight and imagination to adapt and seize the moment properly. America is no different. In 1897, the famed officer and strategist Alfred Thayer Mahan took note of the last time the United States faced such a challenge—with the inception of what is now the U.S. Navy. He wrote: "Those who hold that our political interests are confined to matters within our borders, and are unwilling to admit that circumstances may compel us in the future to political action without them, look with dislike and suspicion upon the growth of a body [the navy] whose very existence indicates that nations have international duties as well as international rights, and that international complications will arise from which we can no more escape than the states which have preceded us in history or those contemporary with us" (Mahan 1940). Mahan warned that the high seas had increasingly opened new vistas for commerce and communication. Therefore, the nation that invests in new sea power capabilities would inevitably dominate the globe. The ingenuity and power of the aircraft carrier subsequently fulfilled Mahan's prediction, ushering in an era of American maritime—and ultimately global—dominance. The United States faces the same need to innovate again today. For policymakers, this imperative presents simple yet weighty choices: Will America lead in space, where it can create and facilitate a new economic revolution, bolster the democratic international order, and dominate the next great battlefield? Or will it cede that advantage to others, with potentially ruinous consequences for American primacy and global stability? Whether Washington likes it or not, a scramble for space is inevitable, and in fact, is already well under way. Today, both Russia and China have surpassed the United States in the military space sector and the development of civilian space. Their innovations include China's proposed work in space-based solar power (SBSP) and testing of anti-satellite weapons, and Russia's advancement of hypersonic missiles. China intends to build space vessels that utilize nuclear propulsion, colonize the Moon, and potentially create areas of anti-access and area denial in space. This activity belies the geopolitical imperative of primacy, now playing out in a new strategic domain. Notably, Russia and China have been quite open about their ambitions. Both countries have recognized that nations that dominate space will dominate the globe. These nations are now angling for space dominance and for good reason. The civilization that is the first to establish a durable presence in space will have the most vibrant and dynamic economy; advanced, high-paying jobs; and a technological edge that is second to none. Moreover, the potential for adversaries to put offensive weapons in space will blunt current American military superiority. U.S. aircraft carriers and land-based missiles will become convenient targets. China or Russia's ability to dominate either energy or communication will make other nations into technological vassal states. As such, nothing short of America's current superpower status is at stake. For the United States to maintain its position of primacy, the country must embrace a reinvigorated space strategy. America will need to progress beyond a mere space program and lead a new military, economic, and scientific revolution that will determine humankind's destiny. The stakes here are high; the nation that achieves space dominance will win future military conflicts. The 5,000-year evolution and history of military technology have confirmed this trajectory.

# CHAPTER 7

# Strategic Competition, Great Powers, Threats, and Enemies

The creation of the U.S. Space Force did not occur in a vacuum. The Space Force was established because of growing acknowledgment of rising threats in space, a domain that is increasingly vital to our national security and economic interests (Cahan and Sadat 2021; Office of the Press Secretary 2019). The 100th anniversary of WWI in 2018 forced the United States to reflect on the international conditions that created that conflict and of what the Space Force may face as we march toward the middle of the twenty-first century. The coming decades will be dominated by great power conflict on Earth and in space. Beyond the United States, the three great powers affecting the strategic climate in space are Russia, China, and India. Two of them are clear adversaries of the United States and its allies. India is a wild card, especially considering its twentieth century ties to the Soviet Union and now Russia. All of them have weaponized space. Each takes their historical and terrestrial-based grand strategies and projects them into space. They are all pushing the ideas of geopolitics by people like Halford Mackinder (Mackinder 1904) into space.

Each September, we observe the anniversary of the 9/11 attacks, where non-state actors killed nearly 3,000 people (National Commission on Terrorist Attacks upon the United States 2004). One should wonder what the total deaths would have been had a medium-size power like Iran or North Korea carried out the attacks. Would it be 30,000? Would an attack from a great power be closer to 30 million? As it came to be known, World War I resulted in the deaths of millions (Royde-Smith n.d.). The aftermath of this conflict, failure of the Treaty of Versailles and the League of Nations, rising nationalism, militarism, expansionism among hostile powers, and retrenchment among Allied forces arguably laid the foundation for World War II. The war resulted in the deaths of some 60 million people worldwide, including over 400,000 American military personnel (National WW2 Museum n.d.; O'Neill 2022). The historical lesson of these conflicts is clear: the United States must maintain its military primacy to deter adversaries.

General Charles Brown, the new Air Force chief of staff, recently stated that we could incur WWII level casualties in a war against Russia or China and noted that we must "accelerate change or lose" (Brown 2020).

Into this environment came the U.S. Space Force.

The U.S. Space Force's biggest challenge will be to compete strategically with the other great powers and deter aggression. This is not to ignore medium-power threats, some of them quite serious, including an Iran or North Korea action to set off a high-altitude nuclear detonation (Garretson 2021).

The Space Capstone publication *Spacepower*, which for the first time articulated Space Force doctrine, summed up the new service's main challenge: "The U.S. must adapt its national security space organizations, doctrine, and capabilities to deter and defeat aggression and protect national interests in space" (U.S. Space Force 2020). The document also cites the late Air Force General Bernard Schriever, who notably stated in 1957 that "our safety as a nation may depend upon our achieving space superiority" (U.S. Space Force 2020, 27).

A historical lens should be used to analyze great power conflict, including current and future conflict in space. One way to accomplish this is to study American and allied strategic thinkers such as Alfred Thayer Mahan, Arthur Tedder, Bernard Schriever, B. H. Liddell Hart, Julian Corbett, Baron Jomini Billy Mitchell, Carl von Clausewitz, Halford John Mackinder, and Curtis LeMay. Equally important are those that come out of traditions that our potential adversaries produced like Sergey Gorshkov, Vo Nguyen Giap, Giulio Douhet, Mikhail Kutuzov, Georgy Zhukov, Sun Tzu, Han Fei, and Zhuge Liang. Space strategists should use these thinkers to extend those strategies into space and anticipate adversarial moves in space. Thinkers like Mahan and Corbett can broaden out into the space domain by creating a new "navalism" for American space strategy. Adversarial thinkers like Gorshkov and Sun Tzu give long-term strategic insight into Russian and Chinese plans and mindsets.

Great power is defined as a nation, rather than a state, with global reach and scale. It influences the international relations (IR) system as a whole, can exert hard power and aspects of soft power, and go beyond DIME (diplomatic/informational/military/economic) instruments of power to include cultural and religious influence.

The fundamental theoretical model for this article is the period of the late nineteenth and early twentieth century culminating in WWI, as this was the last time the world faced a proper multipolar system where the United States was absent from world leadership. In this case, most of the actors operated primarily under IR realism with an aspect of democratic realism and moral realism in the Allies' case (Korab-Karpowicz 2010).

The multipolarity of this era from 1871 onward created a bloodbath. In 1914, the rise of Italy and Germany as nation-states ensured that alliance politics, arms races, imperial maneuvering, expansion, navalism, resource scarcity, cultural divides, and political philosophy would collide. Germany, Italy, Great Britain, France, Russia, Austria-Hungary, and Japan used the world as a chessboard. The United States was hemispheric-centric, still dealing with the aftermath of the Civil War, and focused on Western and industrial expansion. As a result, America was not a significant factor at the beginning of WWI interplay. The core circumstance that caused the Great War was the absence of a world hegemon. The Pax Britannica was over, although Great Britain did not yet know it, and the Pax Americana had yet to begin. The medium and lesser powers attempted to use the great powers for their reasons; Belgium, Holland, Spain, China, Serbia, and Turkey all had their part to play in the conflagration. However, at the heart of the multifaceted circumstances that caused the Great War was the absence of a world hegemon.

We may see a replay of this with nations today like Luxembourg and the United Arab Emirates regarding space. Both of these nations are ramping up space activity and may attempt to drive specific space narratives or be exploited by other countries for their own purposes.

Following the end of the Cold War, some IR and foreign policy scholars, such as Francis Fukuyama in his famous work *The End of History and the Last Man*, argued that great power conflict was a relic of the past and that liberal democracy would continue to flourish (Fukuyama 1992). President Obama similarly argued during his presidency that great power conflict is passé, and that the United States should prioritize multilateral issues such as terrorism, climate change, nuclear proliferation, pandemics, energy, and

Table 7.1 Nuclear State Military Spending

| Country | Military Spending (in Billions) | Nuclear Power? |
|---|---|---|
| China | $250.0 | Yes |
| India | $66.5 | Yes |
| Russia | $61.0 | Yes |
| Japan | $47.0 | No |
| South Korea | $43.0 | No |
| Taiwan | $10.5 | No |
| Vietnam | $5.5 | No |
| The Philippines | $3.8 | No |

*Source:* Data from D. L. Davis 2020.

migration (Rothman 2014; The Obama White House 2015). Unfortunately, however, today's global flashpoints are great-power motivated. The list includes the Euro-Russian frontier, the Baltics, the South China Sea, the Korean Peninsula, the Sea of Japan, the Indian Ocean, the Sino-Indian Border, the Taiwan and Korea/Tsushima straits, and the Middle East, specifically Syria and Iraq.

Space could potentially intensify and amplify these flashpoints until space itself becomes the ultimate flashpoint. Dr. Joel Mozer classifies the threats by time. Today it is China and Russia. Tomorrow it might be coalitions of rising space nations and nongovernmental enterprises. In the future, space criminals (Mozer 2021). The DOD's 2020 Space Strategy states: "In particular, China and Russia present the greatest strategic threat due to their development, testing, and deployment of counter-space capabilities and their associated military doctrine for employment in conflict extending to space. In a recent 2022 survey of 500 space professionals, China, Russia, North Korea, Iran, and India were the top five nations threatening the United States and its allies in space (Mehta 2022). China and Russia each have weaponized space as a means to reduce U.S. and allied military effectiveness and challenge our freedom of operation in space." It acknowledges the necessary and obvious. "Great power competition defines the strategic environment" (Department of Defense 2020a). The threat posed by Russia and China ranges the entire spectrum of threats. "Everything from reversible jamming to kinetic destruction. On the reversible jamming side, it's not that hard to do, and there's more actors in that. But the more you ascend that threat spectrum, it becomes really clear that Russia and China are the threats. . . . And first of all, they're building their own space capabilities for their own use. If deterrence were to fail, we are going to face an adversary that has the same set of capabilities that we've employed largely, over several decades. Furthermore, they are doing rapidly developing capabilities and operationalizing them to negate our access to space. They are doing this in a kinetic multi-domain" (Raymond 2021).

The changes today are alarming. The first change is the United States' slow disengagement from the dominating role it has played after WWII, marked by a roller-coaster of lowering or increasing its defense spending and commitments (macrotrends.net n.d.). America has considered retreating from its role as the rules-based liberal international order leader. The fringes of the two major political parties, for different reasons, call on the United States to have either a light or nonexistent footprint across much of the globe.

The slow American withdrawal coincides with the second change. The current great powers such as Russia, China, and India are reevaluating, amplifying, or changing aspects of their grand strategy in a way that resembles a similar reshuffling that took place in the late nineteenth century.

Third, there are ominous parallels between the cauldron that created conflict leading to WWI and those simmering today. China, playing the role of nineteenth-century Germany, seems determined to upset the economic and military stability created by the United States and Japan, especially in naval power and power projection. Neither an adversary nor a wild card, allied Japan is playing the role of the United Kingdom, an old power clinging to its power base by mobilizing nationalism and militarism. Russia, attempting to resurrect its glory by aggressive action, reminds us of a turn-of-the-century France. India, coming on the world stage for the first time yet not quite ready for a prominent role, is reminiscent of the newly unified Italian peninsula of 1861.

As has been noted earlier, especially in chapter 2, space has been permanently militarized. Is space war inevitable? We hope it is not. There are many camps on this subject, ranging from those who believe that as space is the "ultimate high ground," war is inevitable, just as further weaponization is inevitable. Others, less realistic, suppose there is a formula for limiting weaponization, and those engaging in magical thinking push the idea of space as a sanctuary from all of Earth's troubles—many of them believing that international agreements will reduce the chance for conflict (Raju 2021). Clearly, geopolitical history teaches that conflict in space is unfortunately unavoidable. The strategic space environment is already congested, contested, and competitive (Johnson-Freese 2007, 26). As General John Hyten, vice chairman of the Joint Chiefs, stated, conflict is inevitable (Hyten 2002).

Too much could be made of such parallels. Still, the mix is correct: an increasing multipolarity in which new powers rise. In contrast, old powers try to hold on, and alliance systems that ever more constrain actions and decision-making.

## RUSSIA RESURGENT

The world is currently witnessing the contemporary iteration of Russia's resurgence as it prepares for the conquest of Ukraine. Russia has announced it will only be satisfied if NATO removes troops from America's Eastern European allies, shuts out nations like Ukraine and Georgia, and emasculates itself before the Russian Bear. Russia's most recent foray into space involved it blowing up one of its satellites in November 2021 to demonstrate Russian military power in space (Center for Strategic & International Studies n.d.). Russia ranks second in the Global Firepower 2021 Military Strength Index, and according to the International Institute for Strategic Studies (IISS), ranks fourth globally in terms of defense spending, with a defense budget of $61 billion in 2019 (Béraud-Sudreau 2020; Global Firepower 2021). Russia also spent nearly $4.2 billion on space programs in 2018 (Seminari 2019). A Russian Ministry of Defense document requested by President Vladimir Putin in 2003 spoke in terms of Russia's geopolitical realities with the transatlantic alliance to its west, the Islamic world to its south, and the Asian-Pacific world to its east (Blank 2011). Russia's armed forces must have sufficient power to deal with the challenges posed on each of these axes. Four years later, the Russian media announced the declaration of the Putin Doctrine (President of Russia 2007). Putin himself said that Russia viewed the United States and NATO's policies as threats to Russian national interests. He particularly called attention to NATO's expansion. Russia, therefore, views these as intrinsic threats to its vital and national interests. He warned that deploying a U.S. antiballistic missile system into Eastern Europe would be an abrupt step toward a new arms race. Later, Russia endorsed using energy as part of coercive diplomacy and the old Soviet method of using arms control and reduction agreements to achieve Russian national interest. The country is building and "demonstrating a spectrum of capabilities and resolve to use military force" (Klein 2006).

Throughout all these documents and statements, Russia needs to be treated with the respect granted to the old Soviet Union. Putin achieved this when he forced President Obama's hand over the chemical weapons issue in Syria (Hisham 2017). Russia became the proactive indispensable nation for a false road to peace. In reality, Russia could outmaneuver the United States on the diplomatic front and throw a lifeline to Syria's Assad regime in one fell coup de grâce. In Syria, Russian foreign policy outflanked the United States on numerous occasions with an ultimate strategic goal of power projection in the Middle East and the Mediterranean.

Russian grand strategy and its strategic culture had always emphasized access to a warm water port (Delman 2015), a strong defense against the Turks (Reynolds 2019), the use of buffer states (Toucas 2017), and a fear of others, almost an IR paranoia exhibited by its continuous tumult between Westernizers and Slavophiles (globalsecurity.org n.d.b). It was also committed to an expansionistic pan-Slavism. The Putin Doctrine also aims to reassert Russian regional hegemony, and, supported by a rising and nationalistic Orthodox Church (Wallace 2015), Putin has borrowed elements of the nineteenth-century (and even older Romanov imperial dreams) Russian state to justify a return to an imperial path. Russian thinking is often bold and reckless, exemplified by the thinking of its WWII commander Marshall Zhukov. "If we come to a minefield, our infantry attacks exactly as if it were not there."

The Soviet grand strategy was governed by creating and exploiting the "constellation of forces" to benefit the socialist motherland. Russian strategic thinking today is dominated by several factors, all of which make a window into their quest for space power: the border it shares with Eastern Europe, NATO expansion, its border with China, a blessing and curse of natural resources, military modernization, nuclear weapons, and national pride. One of its greatest fears is an attack along its periphery. This requires the creation of buffers between itself and potential adversaries. Russia can do this by claiming to protect ethnic Russians in what it often calls the "near abroad," where Russian minorities are large and loyal to Moscow (McCauley and Lieven n.d.). One can postulate that this strategic thinking and fear will precede Russia into space. It is also worth noting that Russian thinking is very practical and will likely be so in space. They don't need to have the best technology to win. The godfather of Soviet strategic naval thinking, Admiral Sergei Gorshkov, who turned the Soviet navy into a blue-water power, had a famous quote, "Better is the enemy of good enough," that should give Western strategists pause.

Crimea, Georgia, and to a slightly lesser degree Ukraine and Moldova offer places where Russia can establish "breathing space" from the Europeans; the Caucasus from the Turks and Iranians; and Central Asia from the Chinese (Gabuev 2018). Belarus is in a class by itself, as it will form a joint defense system that will legitimate larger concentrations of Russian troops on the Polish and Baltic frontier (Jacobs 2020). Russian interference in the domestic politics of Macedonia and Montenegro, a severe interest in gray-zone warfare where Russian mercenaries and combatants are involved as far afield as Libya, and their attempt to reestablish Syria as a client state give a window into Russian intentions. Gray-zone warfare is likely to concern the United States in space. It certainly gives new meaning to the often-used gray-zone moniker of "little green men." Gray-zone warfare is a conflict that falls in the usual spectrum between peace and conventional war (Kapusta 2015). It is often the preview of non-state soldiers (often false "volunteers" of national armies), proxies, economic coercion, lawfare, and cyberattacks. As space is akin to a new Klondike gold rush, it is likely to see gray-zone proliferation, especially Russia (Malachowski 2021).

The Putin regime announced a new military modernization program that runs through 2025, with a proposed injection of $770 billion over the next 10 years (Stratfor 2014). A future force will be smaller but more capable of handling a range of contingencies on Russia's periphery. Its priorities for the strategic nuclear forces include force modernization, the development of hypersonic missiles, and command-and-control facilities upgrades (Woolf 2020). Russia will field more road-mobile SS-27 Mod-2 ICBMs with multiple independently targetable reentry vehicles (Woolf 2020). It also will continue the development of the RS-26 ICBM, the Dolgorukiy ballistic missile submarine and SS-NX-32 Bulava submarine-launched ballistic missile, and next-generation cruise missiles (CSIS Missile Defense Project 2021). Russia's Black Sea Fleet will receive 30 new ships by 2020 and become self-sufficient with its infrastructure in the Crimean Peninsula (RT 2014).

Russian nuclear doctrine under Putin has evolved to where nuclear weapons use based on quick escalation is not unthinkable (Ministry of Foreign Affairs of the Russian Federation 2020). Defense experts refer to this as the "escalate to de-escalate" doctrine. Russia would potentially use nonstrategic nuclear weapons in the event of a conflict with NATO to gain a battlefield advantage (Woolf 2021). The 2018 Nuclear Posture Review acknowledged that Russia is expanding and modernizing its arsenal of nonstrategic nuclear weapons (Department of Defense 2018).

Russian space strategy reflects its current and historical grand strategy. The United States and its allies (such as NATO, South Korea, Japan, and ANZUS) are mono-obsessed with China. This is a mistake for many reasons. First, China is the most severe threat to allied geopolitical interests, but that is different from dismissing the Russian Bear. Second, some national security experts still see Russia as a more severe threat (Woolsey 2021). Third, Russia has a history of deprivation, setback, disaster, and incompetence. Marshall Mikhail Kutuzov's famous quote about the Napoleonic war is prescient: "Napoleon is a torrent which as yet we are unable to stem. Moscow will be the sponge that will suck him dry."

Yet one should take to heart Edward Luttwak's famous quote from the Cold War, "Drunk they defeated Napoleon, and drunk again they defeated Hitler's armies and advanced all the way to Berlin" (Luttwak 1985, 230). Drunk, they could win against NATO.

President Vladimir Putin has reignited the previous space program that was in decline following the collapse of the Soviet Union. Citing threats from U.S. missile defenses and programs like the X-37B experimental spaceship, Putin restarted various counter space programs to prevent Russia from falling behind (Ellyatt 2019). Russia will also likely deploy new anti-satellite weapons within the next few years to threaten U.S. space assets (Harrison, Johnson, and Roberts 2019). We are witnessing the era of the rebirth of Russia's counter space strategy (Klein 2006, 100).

Putin's statement can describe Russian's intentions in space: "It is necessary to drastically improve the quality and reliability of space and launch vehicles . . . to preserve Russia's increasingly threatened leadership in space" (Henry 2018).

Space access and denial are critical components of Russian space strategy, essential to modern warfare.

Russia will continue to modernize its military, use covert operations, and economic intimidation to neutralize or co-opt borderland areas while attempting to project power abroad. This will, in part, be focused on blunting America's ability to act unilaterally or to create and hold together alliance structures and coalitions.

Another window into Russian space doctrine is the country's adventurism and military expansion into the Arctic, clearly a hard-power push toward domination (Gricius 2020).

Like all the great powers, Russia is at the genesis of creating a space force. The Russian Aerospace Forces is in many ways a three-branch service combining elements of the space forces, air forces, as well as air and missile defense forces under a single command.

The Russians are developing enhanced jamming and cyberspace capability as well as advanced weaponry such as directed energy weapons, on-orbit abilities, and ground-based anti-satellite missiles to achieve a range of reversible to nonreversible effects. Russia heavily invested in and planned to deploy both FOBS and MOBS nuclear weapon capability during the 1960s (Klein 2006, 99). These were initially designed to stay within the limits of the Outer Space Treaty. "A space weapon system can be deployed in various modes. It might be used as a Fractional Orbit Bombardment System (FOBS), in which the vehicle is placed into orbit, but before it completes one revolution around the Earth, a warhead is reentered on a pre-targeted facility. A more complex mode is the Multiple Orbit Bombardment System (MOBS). A MOBS would be placed in orbit for varied periods of time; then eventually the warhead would be reentered on a pre-targeted or even retargeted facility" (globalsecurity.org n.d.a). The service will monitor space objects and identify potential threats, attack prevention, and carry out spacecraft launches and place into orbit controlling satellite systems.

This has not gone unnoticed by the United States. Earlier this year, General John "Jay" Raymond, the chief of space operations of the U.S. Space Force, detailed how Russian satellites were tailing American spy satellites (Hennigan 2020).

A more significant strategic concern is Russia's plans to establish a Moon colony between 2025 and 2040 (Roscosmos 2018). Russia also recently signed a memorandum of understanding (MOU) with China to jointly construct a lunar research station on the Moon's surface or in lunar orbit. In addition, it plans a joint robotic asteroid mission in 2024, which may mean Russia's exit from the International Space Station (Kramer and Myers 2021).

The current Russian space doctrine can be titled the 3 D's: Disparate, Desperate, and Dynamic. The establishment, fall, and rise of their independent GPS, called GLONASS, is an excellent example of all of this.

One cannot determine when experts dismissed Russia, yet its resilience and willingness to endure deprivation and long-term sacrifice spoil this myopic view. As the Western world obsesses over China, Russia may rise to be the longer-term threat, or at least a threat the West will pay a high price for ignoring.

## THE DRAGON REBORN: CHINA

China's strategic doctrine since the Deng Xiaoping era has been defined by the phrase "to preserve China's independence, sovereignty, and territorial integrity" (Ministry of Foreign Affairs, the People's Republic of China n.d.). In recent years, other slogans and statements have been added, such as desiring a "harmonious world" system and taking advantage of a period of "strategic opportunity" (Poole 2014). Over the last decade, the United States has lost many war game simulations to China (Broad 2021). Chinese actions are the number one concern of the United States in space. The U.S. Department of Defense clearly recognizes this as an enduring threat as both Russia and China weaponize space and prepare for anti-access and denial strategies (Department of Defense 2020b). China acknowledges that dominance in space will decide the victorious factor of future conflicts (Klein 2006, 97). Space is their lynchpin to achieve their global empire (Carlson 2020, 9). China wishes to seize the wealth of the NewSpace economy before the West can react (C. Smith 2020).

China ranks third in the Global Firepower 2021 Military Strength Index, and according to the IISS, it ranks second with a defense budget totaling $181 billion. In addition, China's space budget is estimated to be around $8 billion (Béraud-Sudreau 2020; Campbell 2019; Global Firepower 2021).

The Mao Zedong era attempted to destroy the "olds" of Chinese Taoism, Buddhism, Christianity, and classical Confucianism. As a result, China is filled with bellicose nationalism and wounded pride (Colarossi 2020). The Chinese Communist Party (CCP) and its allies in the People's Liberation Army (PLA) use aggressive nationalism to unify the Chinese people. Yet this use of the rage of the Chinese population is a tiger that, once unleashed, is difficult to put back in its cage. This is illustrated by its inability to control the reaction to government-inspired protests after the accidental bombing of the Chinese embassy in Serbia in 1999, and each time a new history textbook comes out in Japan. As a result, China's potential civil disorder problem is more significant than that of any other great power. There is no difference between the party, the government, and the large Chinese business enterprises. They work in unison and collusion.

Influenced by ancient Chinese strategists like Sun Tzu, famous for many aphorisms, especially "winning without fighting," or Zhuge Liang's "Do the unexpected, attack the unprepared," Chinese strategy has always been based on using diplomacy when in a weak position, deception, and force when it was able. China has never forgotten the period they call the "era of national humiliation" during the nineteenth and twentieth centuries and wished to rise above those who they believe caused it. The current Chinese government is even more indebted to Han Fei, the practitioner of amoral statecraft. Under President Xi Jinping, China has resurrected neo-Maoist evangelism and appealed to third-world Marxists. Xi's ideology is anti-democratic, self-righteous, and revanchist. In many ways, China is restoring Ming and Qing dynasty ambitions by trying (with much difficulty) to create semi-vassal states in Burma, Thailand, Vietnam, and North Korea. If the battle in Russia was and is between Slavophiles and Westernizers, the struggle in China was and is between the "Yangtze River" mentality of sitting behind the Great Wall like the late Ming and Mao period and the "Pacific Ocean" view of domination through adoption and expansion exhibited by Zheng He's treasure fleet and the current President Xi Jinping. Xi is just as pivotal a personality as Putin. Xi's "dream" is that China dominate space, beginning with cislunar domination. Nothing could be more evident than this latter view concerning the space front.

This "Dragon Reborn" creed has been the longest-term threat to peace in the Pacific since the 1930s. The most worrying aspect of this doctrine is China's attempt to engage in the twin plans of building a blue-water navy (Erickson and Collins 2012) and expanding its capability in anti-access/area denial (A2/AD) weapons and tactics (Missile Defense Advocacy Alliance 2018). Driven by resource instability, nationalism, and jingoism, this "First and Second Island Chain" policy envisions Beijing's ability eventually to neutralize or push out America's bases and aircraft carrier fleets, followed by regional dominance.

All of this is expressed in concrete actions: the creation of a significant naval base on Hainan island (Lendon 2020), a massive increase in land-to-sea ballistic missiles (Missile Defense Advocacy Alliance 2018), enormous investment in modernizing China's strategic nuclear arsenal (Center for Strategic & International Studies China Power 2020b), the deployment of its first aircraft carrier (Center for Strategic & International Studies China Power 2020a), the development of its first nuclear-powered ballistic missile submarine (Roblin 2020), an immense investment in offensive cyberwarfare operations (and attacks) (Cybersecurity and Infrastructure Security Agency n.d.), intimidation of Hong Kong (U.S. Department of State 2019), diplomatic isolation of Taiwan (while offering economic carrots) (Sacks 2019), arms and missile technology proliferation (Arms Control Association 2017), anti-satellite missiles (Defense Intelligence Agency 2019), space weapon research (Defense Intelligence Agency 2019), use of the North Korean regime as a bargaining chip (Tkacik Jr. 2005), development of naval-friendly places in the Indian Ocean (Stanzel 2019), and attempting to create a Chinese Air Defense Identification Zone over the Japanese Senkaku Islands (Axe 2020). China is ahead on shipbuilding, land-based conventional ballistic and cruise missiles, and integrated air defense systems (Department of Defense 2020).

Turning from internal Maoism to imperial expansionism, China has had notable success on the world stage, but the flashpoint scenarios it faces are numerous: conflict with Vietnam over the Spratly Islands, conflict on the Korean Peninsula because of Pyongyang's nuclear arsenal, a continuing potential for a confrontation with Taiwan.

China's Belt and Road Initiative (BRI) is most ambitious, encompassing 70 nations (Kliman 2019). It is a global attempt to foster political and economic influence with governments by investing massive amounts of money in infrastructure projects. But, unfortunately, the grease for the engine of the BRI is predatory loan practices and dominance of donated or sold technology.

Many examples illustrate recent Chinese aggressive behavior—a good one is Taiwan. Taiwan continues to be a point of hostile convergence between American and Chinese interests. China is no longer asking whether the United States will come to Taiwan's aid, but even if the United States did, does it even possess the capability to defeat China in the Taiwan Strait?

China's military modernization aims to basically complete military modernization by 2035 and transform the PLA into a world-class military by the end of 2049. China is also expected to double its arsenal of nuclear warheads over the next decade and develop a true nuclear triad (Department of Defense 2020a).

In de jure, China maintains its no-first-use (NFU) policy, but in de facto, it has begun to stray from its NFU policy (Satherly 2020).

There is also possible cooperation and coalition-building with Russia against the United States in places like the Arctic (Cammarata 2020). Further, China controls about one-third of all rare earth metal reserves and owns 80 percent of the global market exports (Li 2019). This monopoly on rare earth metals makes China a key player in the future competition for critical resources.

Space is critical for China as a great power and potential hegemon. Theirs is a grand strategy that includes territorial claims, lawfare, and dominating space economics, particularly production, trade, and energy (Carlson 2020, 71). China wishes to be the dominant space power by the hundredth anniversary of the establishment of the PRC in 2049 (Goswami 2019). As General Kwast describes it, "Communist China has proven itself to be lawless, deceptive, and intolerant in their opportunistic quest to become the dominant world power by 2049, which marks the hundredth anniversary of the communist revolution" (Carlson 2020, 71).

Giulio Prisco argued in a recent article that China wants "cislunar space supremacy" (Prisco 2020). Many quoted Ye Peijian, the Chinese lunar exploration program head, stating, "the universe is an ocean, the moon is the Diaoyu Islands, Mars is Huangyan Island. If we don't go there now even though we're capable of doing so, then we will be blamed by our descendants. If others go there, then they will take over, and you won't be able to go even if you want to. This is reason enough" (Davis 2018). China is obsessed with "first presence" and currently exhibits the world's second-largest space budget at $8 billion per year (Campbell 2019).

China wants complete spectrum dominance, including artificial intelligence, quantum communications, domain awareness, satellite swarms, propellants, resiliency, and intelligence. China's goals are to visit Mars; send probes to asteroids, Jupiter, and Uranus; develop quantum satellites; build a scientific research station in the Moon's southern polar region; and establish a sophisticated large-scale space station within 10 years (U.S.-China Economic and Security Review Commission 2019, 80–90). In 2019, the PRC continued developing its space launch capabilities, providing cost savings through efficiency and reliability, extending their reach into multiple Earth orbits, and improving their capacity to reconstitute space capabilities in low Earth orbit rapidly (Department of Defense 2020a) In 2020, China reached total operating capacity with the BeiDou-3 worldwide constellation, providing mass positioning, navigation, and timing capabilities to its users and additional command and control for the PLA, reducing its dependence on U.S. GPS. However, we know that China is already developing warheads to smash U.S. satellites and lasers to damage sensors.

China plans to place a permanently operating space station by 2022. By 2025, it plans to construct a lunar research station that will evolve into an established crewed lunar research and development base by 2050 (U.S.-China Economic and Security Review Commission 2019a, 90). They are using a similar timeline to pursue space-based solar power (U.S.-China Economic and Security Review Commission 2019, 76–89).

Under the current scheduling, China will be the following country after the United States to send an astronaut to the moon by 2030 (Wall 2019) and is pursuing the establishment of a Mars base, the prototype of which they are currently testing on Earth (Peter 2019).

China's privatized space industry is flourishing, as are the private–military partnerships. The China Aerospace Science and Technology Corporation states that China plans to become the most developed space power by 2045 (Goswami 2019).

China's development of a space force is beyond that of the other great powers. The Chinese equivalent of the Space Force has identified space as a significant vulnerability for the United States and is doing everything it can to capitalize on that vulnerability by advancing its space capabilities as quickly as possible. Creating the People's Liberation Army Strategic Support Force (PLASSF) in 2015/2016 made one organization responsible for developing the PLA's space and information warfare forces. This will allow China to integrate its capabilities into a space force to better play an essential role in future conflicts by enabling long-range precision strikes and denying other militaries the use of an overhead command, control, communications, computer intelligence, surveillance, and so on reconnaissance systems (Department of Defense 2020b). The PLASSF is a combat force to project military power and combat capability into space. It could likely be the number one enemy of the USSF.

The PRC continues to strengthen military space capabilities despite a propaganda public stance against the weaponization of space (Defense Intelligence Agency 2019).

China claims to be building a "nuclear fleet" of carrier rockets (Thompson 2017). Reusable hybrid-power carriers will be ready for "regular, large-scale" interplanetary flights and carrying out commercial exploration and exploitation of natural resources by the mid-2040s (Thompson 2017). According to state media, they will have the ability to mine resources from asteroids and build solar-power plants in space soon after (U.S.-China Economic and Security Review Commission 2019, 88–89).

"The nuclear vessels are built to colonize the solar system and beyond," Wang Changhui, associate professor of aerospace propulsion at the School of Astronautics at Beihang University in Beijing, stated (Rosenbaum and Donovan 2019; U.S.-China Economic and Security Review Commission 2019).

If Russia is the 3 D's, China is the 3 A's: Adventurous, Advanced, and Aggressive. The new American presidential administration is reconsidering its options about China, including an aggressive containment strategy. Space will likely be the arena where the battle between American and Chinese dominance will be decided.

## INDIA: EXPANDING POWER, WILD CARD, OR REGIONAL HOSTAGE?

The newest great power is India, although it has yet to define its place worldwide. It is continuously sucked into the vortex of radical Islam and subcontinental geopolitics that have prevented it from devoting energy to great power maneuvers. India spent much of its postindependence history as nominal leader of the nonaligned movement and has dedicated minimal attention to a grand strategy. India ranks fourth in the Global Firepower 2021 Military Strength Index, and according to the IISS, ranks fifth with a defense budget of $60.5 billion (Béraud-Sudreau 2020; Global Firepower 2021). It is a wild card in the strategic balance with a fierce anti-Chinese foreign policy and a legacy of pro-Soviet/pro-Russian relations. It has also developed a dark and sinister approach to non-Hindu religious groups inside India, notably Christians and Muslims.

India's strategic outlook is often seen in the context of Hinduism and Hindu nationalism, using concepts like Niti (difficult choices, unworthy means to achieve good ends), Artha (prosperity), Dharma (moral obligations, duty), Mandala (geopolitical configuration), and Danda (force and punishment) (Goswami and Garretson 2020, 258).

India faces the rise of Islamic extremism, border tensions with China, perennial strife with Pakistan, and the inability to remedy a level of domestic poverty and corruption that none of the other great powers are burdened with. Moreover, India faces the IR fear of being left out of original foundational groups like the Non-Proliferation Treaty and Missile Technology Control Regime.

Most importantly, it appears to be on a collision course with China dictated by geographical proximity, resource scarcity, and historical enmity.

In 1991, India began its "Look East" policy as an alternative to China in South Asia, targeting Nepal, Burma, Vietnam, Indonesia, and Thailand for greater attention (Haokip 2011). The policy was generally an economic success, but geopolitical assessments have been mixed. India has been unable to become a genuine alternative to China or Japan. As a result, it recently rebranded the policy using the phrase "Act East" (Jaishankar 2019).

Spurred on by Hindu nationalism, Prime Minister Narendra Modi has indicated a new assertiveness in foreign policy, an emphasis on strategic thinking, and a desire for India to become a regional power—all of it supported by a more significant role for the Indian military (Pant 2019). This Hindu nationalism contains a volatile and violent mix that has impacted Indian domestic politics and international affairs.

The much-discussed "Transformation Study" by General V. K. Singh created a window into India's new strategic thinking (Gokhale 2011). It envisions Indian military fighting on "two-and-a-half fronts" (namely, against China, Pakistan, and an Islamic insurgency at home) (Singh 2020). Yet this seems a grandiose ambition because India has been unable to come to terms with a consistent policy over its three major geopolitical issues: Pakistan, China, and the Indian Ocean. Ultimately, India's decision over this last issue will determine its pathway as a great power. A new generation of policymakers has indicated that they want to see the Indian Ocean as an Indian lake. Should New Delhi pursue this vision, it will increase tensions with Beijing. Although India spends only $60.5 billion on defense, its naval trajectory is headed toward power projection. It has two aircraft carriers, and by 2022 intends to have a third (Sharma 2022). This would give it the largest carrier fleet in the Eastern Hemisphere, aside from the United States.

India's problem is how it will build a great power's technological and military capabilities in the absence of a clear goal or strategy. In conceiving of such a strategy, India will ultimately be forced to choose sides, which it has avoided since independence. That choice will dramatically affect the worldwide geopolitical situation.

India faces a fork in the road by being on the cusp of becoming a space power but spending only $1.2 billion on space (Jayaraman 2015). India has been participating in the global space arena primarily focused on making scientific advancements and discoveries on defense or space-based civilizations. This can be seen with the country's Chandrayaan project, which, so far, has sent two probes to the moon (Suri and Guota 2019). India strives to launch its astronauts into space by 2022, becoming just the fourth country to do so behind the United States, China, and Russia (Bartles 2018). It is also increasingly collaborating with the United States on lunar exploration (Howell 2020). Like Russia, GPS is one area in which India is becoming more autonomous with its Indian Regional Navigation System and its

Polar Satellite Launch Vehicle, launching satellites from India, the USA, and Brazil in 2021.

The Indian space force is rudimentary. India's first military application of space was directed at the surveillance of Pakistan. India is forming a space force equivalent to the Defense Space Agency (DSA). In April 2019, India initiated the DSA to command its military space assets, including the military's anti-satellite capability (Rhaguvanshi 2019). The DSA is also charged with formulating a strategy to protect India's interests in space, including addressing space-based threats. The country successfully tested an ASAT weapon in March 2019 (Urrutia 2019).

The Integrated Space Cell, part of the DSA, has been set up to effectively utilize the country's space-based assets for military purposes and defend these assets from a range of threats. India proclaims that it remains committed to the non-weaponization of space. Still, the emergence of offensive counter space systems and anti-satellite weaponry is seen as a new threat that must be countered (Rhaguvanshi 2019).

Indian grand strategy needs cohesion and foundation and attempts to straddle realism with Hindu nationalism. The other great powers will need to decide if India's intentions are ultimately beneficial or challenging to them, especially in the realm of space activity.

## CONCLUSION

Most great powers use realism, but with notable exceptions, such as Pan-Slavism for Russia, neo-Maoism for China, and Hindu nationalism for India. Nevertheless, it is clear that great power conflict is inherent to the IR system, where strategic culture is a product of grand historical strategy and national security policies are a product of both.

Space is the organic extension of great power conflict. All the great powers are engaged in "space force" creation, and power(s) that have a thriving space strategy will, by definition, have a successful grand strategy for the future. Russian and Chinese grand strategy is currently hostile against the allied nations, especially the United States.

WWI offers a good (not perfect) template for multipolar conflict on Earth and in space, where the presence or absence of U.S. primacy is the most critical factor.

It is evident that a multipolar world, absent a United States hegemon, could lead to more chaos and instability on Earth and in space. Yet it will be the U.S. Space Force that can create that very stability.

# CHAPTER 8

# The Service and the Mission

As seen in previous chapters, creating the U.S. Space Force (USSF) was somewhat laborious and full of obstacles. These obstacles have been political, diplomatic, budgetary, and perhaps most significantly based on interservice rivalry. To the detriment of all, these obstacles still exist today.

This would be bad enough in an era of relative peace. It would be catastrophic in time of war. It is disastrous in a renewed period of aggressive great power rivalry.

The knowledge amplifies that both Russia and China realize the role of space for international security. The Soviets came early to this table with a Bolshevik ideology that "strides forth from Earth to conquer the planets and the stars" (Impey 2016, 27).

The stated mission of the USSF is: "The USSF is responsible for organizing, training, and equipping Guardians to conduct global space operations that enhance the way our joint and coalition forces fight, while also offering decision makers military options to achieve national objectives" (U.S. Space Force 2022). As stated earlier, grand strategy depends on national security doctrine, strategy, policy, operations, and tactics. The twenty-first century will see all of this dominated by space power. If the United States wants to ensure its own and its allies' supremacy, it must make space dominance the mission of the Space Force. Space Force will also be needed to maintain free access to outer space (Goswami and Garretson 2020, 47).

A very "inside baseball" debate among space security professionals is how much USSF must "look down" or "look up." In other words, is Space Force's primary mission to protect American interests, and by specific extension, the American terrestrial military on Earth? This is the "geocentric" model for Space Force. Or is Space Force's primary mission beyond the Karman line. Can it do both? Must it choose one over the other? Most space power advocates see it as a service for the stars and not tethered to the Earth. It must focus on the "strategic terrain of space" (Carlson 2020, 7). Many see the synergy between the two views of the ultimate potential of all space forces (including our adversaries) as attaining "maximum power projection as platforms for kinetic or laser energy weapons or with mass-destruction payloads" (Dolman 2002). In essence, one cannot look down without looking up. This false dichotomy will only alienate portions of the electorate, the national security establishment, and our allies. The purpose of the USSF is the exact purpose of the entire military branch, the protection of American civilization, nothing more and certainly nothing less.

## THE SERVICE AND THE DOMAIN

The sixth branch of the military is currently led by General John "Jay" Raymond, the chief of Space Operations and a member of the Joint Chiefs. The long fight over independence continues, though the ideas to make Space Force a subcategory of the Air Force or a merged service with the Air Force (Brooks and Agrawal 2021) seem over. In the first year, USSF achieved great strides, including:

- Chief of Space Operations becomes the eighth member of Joint Chiefs of Staff
- Activation of Space Operations Command and Space Training and Readiness Delta (Provisional)
- Commissioning of first officers from Air Force Academy and Officer Training School directly into USSF
- Initiation of partnerships between USSF and U.S. allied nations and industries
- Start of joint USSF and NOAA Operations of Infrared Weather Satellite
- First renaming of Air Force bases and stations to Space Force: Cape Canaveral Space Force Station, Patrick Space Force Base
- Finalization and announcement that USSF space professionals will be known as Guardians
- Creation of a new National Space Intelligence Center to serve as service intelligence center for Space Force
- First member of USSF to deploy overseas (part of a Marine Expeditionary Unit)
- First launch of an X-37B Orbital Test Vehicle under joint auspices of USSF and the Air Force Space Rapid Capability Office
- The formation of SpaceWERX, a new entity that will focus on technologies for the Space Force
- Publication of Space Force doctrine and planning guidance (The Space Report 2021)

It is clear that to ensure independence, creativity, innovation, boldness, and accountability, the USSF must be completely independent. It also cannot be seen as a complete realm of the Air Force. There are already non-Air Force personnel who have transferred into the service. The Marines have already been talking about a mobility vehicle that could use space to put a combat squad to any location on Earth (Dinerman 2021, 9).

This question of independence and the degree of independence is one of the central debates. General Raymond recognized the need for independence and pointed out that the secretary of the Air Force currently has two services under them. It is further acknowledged that until Space Force expands, the Air Force will be doing some of the heavy bureaucratic lifting. However, he states, "You have to independently develop your people. The force development aspects of the Space Force are all going to be done independently from the Air Force. You have to design the organization. We have completely reorganized, separating from the Air Force, and built our organizational structure. We have to have our own documents. We have written our own independent theory of space power, which might not be 100 percent perfect, but it's good, and it's a good starting point for a war-fighting discussion, as we progress. And you have to have your own independent budget. We have done that. We've taken the dollars that was associated with space out of the Air Force and made our own independent budget out of that. And that budget today represents just two percent of the entire DoD budget. And then finally, we have to design our force structure" (Raymond 2021).

This does not mean independence from the Department of Defense, but independence from the U.S. Air Force (Mozer 2021). Many national security professionals emphasize the need for Space Force to think like a military branch, be a military branch, and not something else (Kehler 2021). If it does not have true independence, it will get absorbed by the other services, not only in the material realm, but also in the realm of culture, thinking, and motivation (Shaw 2021). There is a subset concern here that pilots too heavily dominate the U.S. Air Force itself (C. Smith 2020). General Kwast uses an excellent analogy from WWII. "The reason the Sherman tank was inferior to the German Tiger tank in World War II is because General Pershing and General MacArthur prohibited the tank from being developed independently of the infantry. Every spare dollar went to what the infantry thought was the most important thing, and that was the soldier and the rifle, and what was left over went to the development of the tank. And that's why the Sherman tank had a pop gun for a weapon and its rounds bounced off the front of the tiger and the panzer. And we lost thousands of American patriots because of that lack of understanding of the power of independence and the development of something new. We also saw it in the development of the airplane. The reason we lost more airmen in the European theater than all of the Marines in both the Pacific and the European theater, is because the Army did not let airplanes be built independently. They required it to be under the Signal Corps and under the Army. And the Signal Corps believed that the airplane was meant for seeing and observing things from up high." General Kwast argued that the previous WWI leadership only thought airpower was useful for watching the troops over the horizon in the trench line because they did not believe in independent air power. It was only due to the lessons of World War II, where America came close to losing battles, that the leaders finally came to the understanding that until you have an air force that is independent from the culture of the army, you will never beat your competition (Kwast 2021). Many if not most space power advocates see the question of independence directly linked to whether or not the United States will be the preeminent spacefaring nation (Cooper 2021). Former CIA director Jim Woolsey sums up what many national security experts not directly involved in the day-to-day space debates think when he stated, "I don't think it matters as long as they [DOD] work together and don't mess it up" (Woolsey 2021).

**THE MISSION**

The mission does not need to be searched for. Since the early days of the Cold War, we have seen the visionaries illustrate the needs to today. Later in 1998, Space Command issued an extended range plan with four parts: Space Control, Global Engagement, Full Force Integration, and Global Partnership (Dolman 2002). We also know the practical objectives in both dominance and governance in the near term. These are LEO/GEO orbits, cislunar space, lunar orbits, transit and surface, and the Lagrange points. The fundamental mission is to support American interests and our allies in space (Dinerman 2021). In simpler terms, it is to protect national security through space power.

However, there exists a great ocean of responsibility and obligations that USSF will need to handle. Many of the recommendations of space power advocates like Coyote Smith have been enacted in the last two years. How far we go to "expand the major force program" or seek to be subservient to the outdated Outer Space Treaty (C. Smith 2017) remains to be seen.

General Raymond has addressed the fears by some that USSF will not cooperate with the military branches. He emphasizes a salient point: USSF is an armed service, a military branch with military roles. The newness of USSF creates unique opportunities since it is the first branch that has been established since the Air Force separated from the Army back in 1947. "And what I tell my team is two things. One, we have to be very bold. We've been given a clean sheet of paper and don't just make incremental changes. Go big. And so one risk that we have is that we don't think bold enough and that we don't make the needed changes that we have to. And two, that when we do think bold, that the bureaucracy says "no, that's not the way we do business" (Adde 2022; Raymond 2021). General Raymond clearly understands the crossroads situation that the new service faces. It must be bold and innovative.

Space Force will be unique in that it will be global and extraterrestrial at the same time. It will also be permanently deployable and more technology dependent. It will possess one of its greatest strengths and weakness in the area of domain awareness and situational awareness. As General Shaw accentuates, what may appear simple is titanic. The space domain is different, and the technologies and the physics of the space domain are different (Shaw 2021). This is not just engaging in Earth activities somewhere else.

One commonality among space power advocates is the need for consistency and cohesion in the mission and the desire for a synchronized whole of government approach to space policy (Gingrich 2021). Many space power advocates are blunt in their assessment of the future mission of the Space Force. In essence, it will be the preeminent military service of the future. Space Force not only will be able to do these things in space, but by the nature of the physical universe, they will be able to do anything they can do in space and the terrestrial realm as well. General Kwast continues by highlighting that whatever the Air Force does, the USSF can do better, faster, and cheaper. The Space Force can be anywhere on planet Earth in minutes applying the capability to protect and preserve an escort or bring to justice anybody that's behaving badly. He warns, "wow, you're 100 years too soon," but people like that don't understand how quickly history accelerates as you go forward. And most people cling to the past or fight the last war and forget about the new force, the new world coming our way. General Kwast fears that the Air Force is building up for the last war, the industrial age-war, and they're "all dead men walking. It would be like building the Maginot Line or trenches for trench warfare in the age of air power or building a moat and a castle to defend yourself, knowing that it's the age of air power. Tanks, ships, and planes in the terrestrial realm, that are not able to go into space and move at Mach speed are literally dead men walking, because of what space will be able to do. So, this gives you a feel for how different the Space Force will be, that if we were to do this efficiently, we would just make the Air Force into the Space Force and start divesting of the ships, tank and planes that defend terrestrial environments, because the Space Force will be able to protect them faster, cheaper and better than any Air Force tool that's being built today" (Kwast 2021). There is clear recognition that, ultimately, Space Force will possess weapon platforms that "will deliver the opening blows in combat operations to knock down enemy defenses, enabling the other services to engage successfully with fewer personnel. In fact, the majority of strikes in war proper can be delivered from space using a combination of lasers and other directed energy systems, in addition to physical kinetic devices. Space will be where we base our weapons because it confers such a positional advantage over terrestrial basing (C. Smith 2020).

The doctrine-level mission is the simplest to declare and the most challenging to implement. It creates a space national security mission held primarily by the USSF to ensure American and allied dominance. This requires a panoply of training, education, and leadership of space national security professionals and soldiers.

Space Force must embrace flexibility, interoperability, and resilience (Dickey 2021). In addition, there are strategic, operational, and tactical missions that the Space Force must embrace.

The strategic missions will be astropolitical. They will include planetary defense, treaty compliance, verification, and finally, a strategic vision to secure U.S. interests beyond GEO (xGeo) and beyond (Buehler et al. 2021). Sometimes this is referred to as "the cislunar area of operations" (Trevithick 2020). It must embrace the virtuous cycle of protecting space commerce and communications as a national security imperative.

The operational missions will be varied, and to do this, USSF must expand beyond Earth, improve its budget and capabilities, build endurance, and be ultimately flexible. Including ISR (intelligence, surveillance, and reconnaissance) and SDA (space domain awareness). Many obligations will be heavily based on a constabulary to forestall criminal, illicit, dangerous, and violent behavior. This could be based on a Coast Guard model (including the creation of space cutters [Ziarnick 2015]), such as inspection, interdiction, boarding, customs, crime investigation, resource and contract protection, and arrests (C. Smith 2020). It will need to garrison the high ground, especially on the Moon. Space Force must support the terrestrial military (brown water) while ensuring a space-based defense of the United States and its allies (green water), and dominate the orbits, LEO and GEO (blue water), and finally cislunar and beyond (black water). Finally, it must possess defensive and offensive weaponry directed at all four layers (colors).

## THE MISSION REFINED—A STRATEGY TO 2040

In November 2021, space professionals and advocates like myself met for a multiday exercise based around potential future scenarios that could show the strengths, weaknesses, and needs of the "Space Force after Next," or how to get to the middle of the century with America as the preeminent spacefaring nation. The reader may remember that we had already tried to position the United States out to 2060 (U.S. Space Command 2019) and were now working backward from that goal.

## EQUIPMENT, TECHNOLOGY, AND TRAINING

Futurists and technologists see the space domain as the ultimate expression of human science, engineering, and ingenuity. As much as space is analogous to the maritime environment, it is harsher and more unforgiving than submarine operations. The challenges of operating in Zero-G, in a vacuum, exposed to electromagnetic radiation, are simply astounding. In the near term, the Space Force will need satellite protection and anti-satellite weapons that allow for kinetic and non-kinetic so that the United States will have layered and multiple options that do not confine our activities. It will require forward-looking, often dubbed "science fiction" thinking to achieve the mission goals of the USSF. There will be the need for new weapons like railguns, directed energy, hypersonic missiles, and space weapons platforms. It will require exploring new fields such as hyper war, "centaur" (the potential combination of human and AI) technology, AI, and quantum computing and communications. Finally, there will need to be a system of logistics and transportation never before seen in any human endeavor.

There will ultimately be the need for a separate Space Force Academy (Dinerman 2021) and individual ROTC programs.

## CHAPTER 9

# Pax Americana and Pax Astra

## EX HADRIANO MURO

Hadrian's Wall was built in the first century CE. Both history and legend serve as physical and psychological icons of the boundary between civilization and barbarians. Here, we should reflect on the Space Force's choice of the word "Guardians" to describe the soldiers of the USSF. The first order of business for any military is to protect the homeland strategically. If they can't accomplish this, they have failed. This is the same for Space Force. They have one real first job, which is to protect American civilization. There should be no talk of colonization, expansion, exploration, transport, logistics, communication, or trade if USSF can't have this as their primary goal. I would be the first to argue that the best way to accomplish this protection is through the dominance of all those variables, as long as we understand that everything needs to be predicated on that fundamental mission. This is put simply in foreign policy terms to protect American vital and national interests. Second, the military needs to establish dominance in its domain; and third, it must preserve and enhance American allies. Finally, it must do all of this with a clear motive to protect the American way of life, culture, politics, and morality along the lines of our Declaration of Independence and Constitution.

There are three guideposts for Space Force to achieve this. These guideposts will lead us to a successful grand strategy. These are the doctrine, the long-term strategy, and the near-term strategy.

## DOCTRINE

U.S. Space Force doctrine must first accept that it is under national security doctrine. Successful American doctrine has always been under nine overarching themes. The first of these is American exceptionalism—the assertion that America is unique among nations across all time. This asserted uniqueness transcends the United States' particular history, geography, or demography. Americans perceive themselves as exceptional insofar as they see theirs as the only nation fundamentally defined by ideology. This American creed or ethos of liberty created a gulf between the United States and all other nations. It has meant that American exceptionalism and American nationalism have always been one and the same. This exceptionalism results from America's inherent liberal character that has often contradicted the dictates of realpolitik. It created a nation diverse in religion but unified in its civic religion expressed in the Declaration of Independence and the Constitution. This "new Jerusalem" was qualitatively different and became a central national security and foreign policy precept. Abraham Lincoln (1862) talked of the American destiny as the "last, best hope of earth," which was "destined to be a barrier against the return of ignorance and barbarism." Similarly, John Adams (1765) noted that America was "destined beyond a doubt to be the greatest power on earth." America is the last best hope for space and humanity's expansion to the stars.

The second is manifest destiny and expansion defined not as popular culture would wrongly do so, but best by senator and historian Albert Beveridge in 1900: "American law, American order, American civilization, and the American flag will plant themselves on shores hitherto bloody and benighted, but by the agencies of God henceforth to be made beautiful and bright." Manifest destiny is a purely American term; it espouses the belief that Americans have been destined to expand as ordained by God in connection to American exceptionalism. Four hundred years of steady expansion were based on this belief. The roots of manifest destiny predate the nation's birth, with the concept of an "imperial republic" taking hold early. "The early colonies were no sooner established in the seventeenth century than expansionist impulses began to register in each of them. Imperial patterns took shape, and before the middle of the eighteenth century, the concept of an empire that would take the whole continent was formed" (Rosati and Scott, 2010). The continued continental expansion occurred alongside the Puritans' desire to create the Christian "city on a hill"—an idyllic and divinely blessed society whose republican government would serve as an example to the world. This "American messianism" created a mission for the United States. During and after the revolution, enlightenment founders justified expansion into Indian Territory to share the blessings of democracy and civilization. As this conflation of national security, foreign policy, and national identity indicates, these state matters and perceptions were completely intertwined. It granted Americans the view that their expansion, culture, and institutions were inherently superior and different from that of their rivals. After the War of 1812 under Presidents Monroe and Adams, the idea became to redeem and remake the world by pursuing a divinely inspired destiny. Manifest destiny combined nationalism, idealism, and self-confidence. American expansion was not for territorial acquisition or great power but as a force of good. It created the twin engine of manifest destiny as a desire to spread liberty and punish evil. Many contemporary scholars have completely corrupted this idea. Thus, it is only fitting that the inspiration for the first moon base and Mars colony are based on human liberty and human dignity principles.

The third is the empire of liberty and democracy promotion. Much contemporary criticism and evaluation have been leveled at President Bush's doctrine and its primary pillar: democracy promotion. However, this promotion was part of America's genesis in connection to manifest destiny. The United States has always dreamed liberal imperial dreams. This theme has received the most robust and consistent reiteration out of the nine. John Adams (1765) recorded this enduringly relevant sentiment: "I always consider the settlement of America with Reverence and Wonder—as the Opening of a grand scene and Design in Providence, for the Illumination of the Ignorant and the Emancipation of the slavish Part of Mankind all over the Earth." America's consistent expression of its values through this goal of nurturing and expanding democracy reached its height in the twentieth and twenty-first centuries. Presidents who deviate from this narrative to prioritize policy undermine American grand strategy, and have generally been ineffectual leaders. This was true of Nixon's doctrine of contraction. In the national security context, approaches based on universalizing American national interest transformed into policies of liberation and rollback in tune with Americanism. This goal of a liberal empire and democracy promotion contribute immensely to America's overall power abroad. This "liberal grand strategy" is a pragmatic approach to American national interests—not merely an attempt to do the right thing. The argument here evokes Emmanuel Kant's "perpetual peace," which forms the basis of democratic peace theory. The approach to grand strategy and national security that champions democracy abroad as a way to secure national interests is patently American and fits within the larger liberal view of international order, civil society, institution building, human rights, free trade, and progress. This point cannot be overemphasized: American promotion of the empire of liberty has never simply been about elections; it has always been about civil society and liberty under law, whose most tremendous success has been in the post–World War II period with Japan and Germany. It will be necessary to establish this empire of liberty extraterrestrially to avoid the other alternatives, which are tyranny and chaos.

The fourth theme is the recognition that free trade, free commerce, and market economics also influence doctrines. This will be fundamental to the NewSpace economic revolution. Economics is structured around the primary American demand for unhindered trade, the free navigation of the seas, and open access to international markets. To achieve this, America has frequently pursued a verifiably practical policy of armed neutrality This idea of armed neutrality began with Washington's proclamation of neutrality of April 22, 1793, all the way to President Wilson's address of April 2, 1917. Neutrality was the official policy of the United States. The United States, from the Founding Fathers onward, viewed any attempt to interfere with American commerce as a hostile act. Its first foreign war was fought against the Barbary pirates partially over this issue. However, protecting goods on the seas fits within a broader belief that America can reform the world through its principles, free trade, and economic influence.

Furthermore, as America grew to be a world power, it recognized that order and trade could not be maintained unless it had complete control of the sea lanes and skies. Thus, the American national interest in prosperity at home could promote prosperity, law, and freedom abroad. In space, America's demand for free navigation and access to and in space as well as ensuring a stabilized economy will be pivotal.

Fifth is unilateralism. The word "unilateralism" has become a contemporary epithet used to bludgeon policymakers as "barbaric." Many scholars have argued that President George W. Bush was unilateral, but no president is truly unilateral when engaged in effective diplomacy and war. We have made the term into a cartoon. However, it is essential to acknowledge that the United States has always functioned within a tradition of unilateralism, which has necessarily replaced the assumed tradition of mythical isolationism. Unilateralism is the oldest of the various realistic traditions in American national security. It exists in connection to the American president as a single power source. This was not simply based on realist notions of power politics and the need to determine one's own fate; part of the belief in exceptionalism demanded American unilateral action apart from the other powers. While acquiring recognition on the world stage, the nation also acquired the need to be unilateral based on worldwide commitments to support the Pax Americana. Unilateralism has allowed America to hold firm to its national morality and goals without automatically adhering to multilateralism. For space, the mission will need to come first. It needs to be a mission based on realism and freedom. Allies who agree with this are welcome.

Sixth is the American adherence to internationalism. Historically, the United States has not shunned its allies or friends; it has often pursued alliances to defeat what it has identified as evil. But again, as America grew in global prominence, it became clear that grand strategy was only sensible within a worldwide context. American internationalism in the twentieth century began by globalizing not just American interests but also American values. This must be the same touchstone in space. Again, in the twentieth century, it became clear that alliances with certain European states would be the source of American national security. This would later be expanded into permanent alliance structures around the world. The recognition that American grand strategy required a national security doctrine enabled the United States to build coalitions of free people to counter the threats of tyranny and extremism. American internationalism, based from the beginning on a commitment to creating alliances when needed, matured along with the nation's sphere of influence. This is the only choice for America leading in space, leading the free people into the stars, and creating collective security while moving outward.

Seventh is the American way of war. The United States was birthed from war—an attribute that distinguishes it from other nations and great powers whose genesis has been rooted in tribe, land, language, and religion. America sprang from an ideology grounded in natural law, and America, as a warlike power, succeeded because of what happened on the battlefield. War is the rule in American history and international experience. Military scholars register 163 military interventions before WWII and some 180 U.S. Marine landings abroad between 1800 and 1934. These operations were punitive, protective, or "profiteering." In the twenty-first century, the United States engaged in wars in Afghanistan, Iraq, Syria, and numerous special operations and low-intensity conflicts. U.S. military objectives in national security are clear: the complete domination of North America by the U.S. Army, the elimination of any threat to the United States by any power in the Western Hemisphere, the complete control of maritime approaches to the United States by the Navy, the total domination of the world's oceans and trading system, and the prevention of any other nation from challenging U.S. global naval power. It is merely amplifying this into LEO, GEO, and cislunar space. The model exists from the beginning of the republic onward. In addition, the American way of war has become total destruction of the enemy—it is a strategy of annihilation—ultimately through advanced firepower in an attempt to minimize casualties and assure total victory. But Americans' primary motivation has not been destruction but to "set in motion America's righteous power" to punish evil and brook no compromise with evildoers. This has generated criticism that Americans embrace crusading military conflict but are uncomfortable with the quagmire of nation-building. However, the ideological component of American war praxis has also led to the formation of history's most humane military. The military has a paradoxical civility that springs from an obsession with upholding honor by intentionally avoiding killing civilians, fully attending to prisoners of war, and strictly adhering to the rules of war. While the question of when to utilize war is at the epicenter of grand strategy, the threat and use of war is clearly the most instrumental tool in American national security doctrine. We will face the exact same questions and answers in space.

Eighth is geography, geostrategy, geopolitics, and what will become astropolitics. The study of geo/astropolitics is the study of what is enduring and draws from that which has existed over centuries and millennia. Geopolitics and astropolitics bring together the demands of a nation and the limits of its geography, and in space, the limits of gravity, orbits, technology, and vacuum. For the United States, only its demands could limit its vast geographical ability to project power. The natural boundaries of two oceans, a weakened power to the north, and a chaotic regime on the southern border greatly influence American national security. Although this was the case for most of its history, geopolitics played a pivotal role in the constitution ratification debates and the years of the early republic precisely because America was existentially threatened by the British, French, and Spanish empires. When America took the world stage, it became clear that geopolitics would dictate its projection of power. Astropolitics will be the new arena, limited only by our imagination, acknowledging the limits and overcoming them.

The ninth and final theme is primacy. As the United States developed, it embraced primacy as a necessity. America's quest to be the sheriff or an order maker began in the nineteenth century. It proceeded apace as the United States grew from great power to a superpower, to the facilitator of the Pax Americana. However, this quest for primacy is not synonymous with the desire for empire. Instead, primacy was sought to ensure that America's grand strategy goals were served. The only way to achieve this certainty was to build a nation without any serious peer competitor for international and military power. This form of primacy simultaneously served American national objectives and interests while increasing democracy, human rights, and free markets.

These nine themes illustrate consistency in American grand strategy. They are exemplified in the most robust and successful national security doctrines, which create successful national security policy and strategy. These will do the same for our space doctrine and strategy. They serve as a historical model, current benchmark, and future standard for analyzing national security, foreign policy, and strategy. Finally, these themes illustrate America's past and the ongoing fight against tyranny. Thomas Jefferson's (1800) quotation applies in this way to the organic nation as a whole: "I have sworn upon the Altar of God eternal hostility against every form of tyranny over the mind of man." This began with the war against the tyranny of monarchy from 1775 to 1918 and was followed by the tyranny of Nazism, fascism, communism, and militarism from 1918 to 1991 and beyond. The United States' current wars are being fought against the tyranny of Islamic extremism and the resurgence or continued existence of worldwide authoritarianism and totalitarianism. In space, America will pledge itself to the same values to ensure that freedom for humanity and protection from tyranny continue. Following these values will guarantee liberty, prosperity, and security as they guide America into the twenty-first century and beyond.

## PROTECTING LIBERTY AND DEFENDING THE EARTH

As much as I am a strong proponent of Space Force's official motto, *Semper Supra* (Always Above), I advocated the following motto at the first Space Futures Workshop—Protecting Liberty Defending the Earth. It is also worth mentioning that Space Force is the military branch, but Space Command is the warfighting component. Thus, the doctrine and strategy are really for both. It also questions whether these combatant command structures work for space. Finally, beyond the current purview of this book, there is a grand question about what roles the other military branches will have or want in space.

Space national security doctrine must follow the American grand strategic goals of preemption, prevention, primacy, and democracy promotion. Preemption is acting on an immediate threat before it hits you, prevention is acting on the clear threat on the horizon before you are forced into the crisis mode of preemption, primacy as already defined allows for no peer competitor, and democracy promotion is the facilitation of liberty, human rights, and free commerce. In short, this is a pathway to "black-water" space policy.

## LONG-TERM DOCTRINE FROM THE MIDDLE TO THE END OF THE TWENTY-FIRST CENTURY

As we head into the mid- to late twenty-first century for space, we know certain parameters must be met. The majority of these parameters must be the responsibility and obligation of the U.S. Space Force in concert with American and allied military branches. The U.S.-led alliance system must accomplish the following.

### United States—Dominant Space Power

As discussed throughout this book, space power and national security will merge to become one and the same. According to the Annual Threat Assessment, which is released by the Office of the Director of National Intelligence, Chinese and Russian space activity ought to be one of the biggest dangers for the United States and its allies (Donovan 2021). Not only do both countries possess the necessary weapons to damage the United States, but it could seriously hurt the military if satellites related to the U.S. military were hit (Office of the Director of National Intelligence 2021).

China seeks to become the dominant space power besides the United States. Currently, they wish to "match or exceed" American capabilities in space. China would gain powerful military, economic, and prestige benefits by closing the gap and eventually passing the United States in space proficiencies. The United States expects a Chinese space station in low Earth orbit (LEO) to be operational in the frame of 2022 to 2024. Furthermore, China is heavily invested in establishing robotic research stations on the Moon and a crewed base. The Threat Assessment also states that the Chinese military is continually integrating space services (e.g., navigation, timing [PNT], satellite reconnaissance and positioning, etc.) into its command-and-control and its weapons systems "to erode the US military's information advantage" (Ibid.). Destructive and nondestructive antisatellite (ASAT) weapons, both ground- and space-based, have also become a significant part of China's space program.

Russia, until recently often ignored in global military calculations, is the other main competitor in space, as they have maintained "a large network" of communications, reconnaissance, and navigation satellites (Ibid.). Like China, the Russian government is continually working on integrating its space services into its command-and-control and weapons systems (Ibid.). The country's primary focus on space is the disruption and degradation of U.S. (and allies') capabilities by fielding new ASAT weapons and training their military space elements (Ibid.). Therefore, the Russian ambitions directly interfere with American space dominance.

In order to stay the world hegemon, the United States must be able to protect itself and become "so resilient, that no matter how deadly the attacks, it will function well enough for the military to project power halfway around the globe in terrestrial reprisals and counterattacks" (Broad 2021). Space is critical on the terrestrial front militarily for targeting, communications, positioning, timing, and location—in all of which the United States has a strategic advantage at the moment due to its current space dominance (Lopez 2019). If it wishes to remain in its current premier position, space power must go far beyond terrestrial support and become its own domain where American space power dominates the system.

### Permanent Norms and Governance of Space

Chapter 5 illustrated the issues of norms and international law. The current global governance for space is not practical in response to space attacks, being only one of the many reasons why governance must be reformed and permanent norms must be established. The United States has publicly stated the necessity of "strengthening global governance of space activities" and has committed to establishing a "rules-based" international order for space and developing new measures to contribute to safety, stability, and security for all actors involved (The White House 2021). They want to establish this by working with its allies, commercial industry, and other partners (The White House 2021).

Space exploration now includes more than 70 countries, not only the United States and Russia anymore, meaning that new norms and space governance will need to entail additional threats faced by all actors.

A recurring issue the United States has run into when establishing permanent norms and governance for space is the different interests of parties involved in space. Since warfighters face a completely different mission than collectors, the U.S. defense and intelligence communities have had significant issues reaching consensus within their sphere (Ohlandt, McClintock, and Flanagan 2021). But, overall, norms should play a role in detecting and responding to potential threats and mitigating space debris and other factors that make the environment more dangerous (Schaffer 2017).

In order to establish a healthy climate in space, one in which the United States will be able to remain the dominant power, U.S. intelligence and defense communities must first reach consensus among each other before conveying their proposals onto an international stage. It must be identified by all parties involved which norms all parties support and what measures all parties want to avoid to keep the United States safe. A treaty on space debris, for example, is an issue that must be tackled on an international level. Though norms and governance will not end all conflict, the current "lack of equivalent norms in space allows actors to operate any location in the domain and at any distance from other spacecraft" (Ohlandt, McClintock, and Flanagan 2021). This lack could lead to severe issues for the United States in the future. And though norms and governance will not be the end to all conflict, it is a helpful tool. The United States must lead this issue, remain the leader, and shape the field of norms and international law for space to fit its strategic and moral values. American leadership is the critical key for this to work.

### Primacy of Cislunar Space and Then Beyond

Cislunar space, often referred to as the "gateway to space," is commonly viewed as a vital area to space supremacy (U.S.-China Economic and Security Review Commission 2019). The entire Earth's orbit can be observed from this area, allowing greater command, which is essential to control on Earth and space. According to Xi Jinping and the PLA, whoever controls cislunar space has "good command and control architecture" (U.S.-China Economic and Security Review Commission 2019). Moreover, cislunar space is vital to obtaining and transporting resources in space, which are fundamental to future space missions that might go past the Moon (Vedda 2018).

One aspect of cislunar space that is imperative for success in future space endeavors is the dominance of the five so-called Lagrange points. In these areas (L1, L2, L3, L4, L5), the gravitational effects of the Moon and Earth are canceled out, allowing for stability (Maher 2020). This would allow whoever has power over those areas to permanently place structures or spacecraft there, displaying an immense advantage (Maher 2020). The importance of these Lagrange points is well described by Captain Bryan

W. Maher of the U.S. Air Force: "Who controls circumterrestrial space could dominate Planet Earth; who controls our Moon could dominate circumterrestrial space; who controls the L4 and L5 points could dominate the Earth-Moon system" (Maher 2020).

Due to the Lagrange points, the possibility of permanent spacecraft or other structures, the possible exploitation of cislunar resources by China, and its strategic importance to space warfare all make it extremely important to first protect cislunar space before branching out further. The United States will need to claim dominance over this space, not for a "present tactical advantage," says strategy consultant Peter Garretson, but rather out of "fear that China's moves to cislunar space will provide it with a positional and logistic advantage from which it could occupy, constrict, threaten or coerce US interests" (Maher 2020).

Cislunar space is the new interior lines of military conflict. It is to American space power what American naval dominance is to the Atlantic and the Pacific Oceans. If America fails to gain the absolute advantage here, the entire space power enterprise will fall apart.

### Astropolitical Control of Strategic Space Locations and Orbits

The most critical space topography in Earth's orbit to all state actors is the gravity well. This area entails the strategic locations and vulnerabilities, such as the most exposed space, which is the low Earth orbit, which the United States must control (Jenkins 2021). Matthew Jenkins from thespacereview.com has identified key space locations and uses standard military terms to explain the areas better (Ibid.).

The first area that Mr. Jenkins identifies is "lines of communication," which is the most important strategic position. All nations—friendly and adversary—currently "utilize all orbital regimes concurrently" and in very close proximity to each other. These lines of communication used in space can be in physical or nonphysical forms. Dr. John Klein defines celestial lines of communication as "those lines of communication in, though, and from space used for the movement of trade, material, supplies, personnel, spacecraft, electromagnetic transmissions, and some military effects." Being the most important space location, this will determine space dominance within this century.

Another geographical term Mr. Jenkins relates to space is the high point/high ground. He identifies that this point gives strategists an optimal view that allows them to plan ideal defensive positions and fields of fire (Ibid.). Mr. Jenkins believes that satellites from high altitudes can keep an eye/ear on Earth, mitigating exploitation and improving communication (Ibid.). Most countries believe the GEO high point to be the area to place intelligence, telecommunication, and other essential satellites to achieve the advantage described above (Jenkins 2021). Lastly, the choke point proves

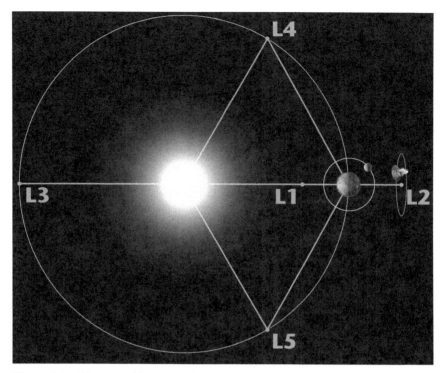

**Figure 9.1** Diagram of Lagrange Points
*Source:* NASA/WMAP Science Team

to be a vital factor for space strategy. This will be an area of transition for space-based assets from LEO to GEO. Dr. John Klein argues that LEO has already reached a very critical level of density and has, therefore, already become a choke point for space systems (Ibid.).

Other strategic space locations vital to the United States' success in space are the Lagrange points, as shown in Figure 9.1. As previously discussed, the Lagrange points in outer space are areas with enhanced attraction and repulsion, allowing for permanent structures and spacecraft. Furthermore, fuel consumption is reduced significantly, allowing for these spacecraft to remain in position for a significant amount of time (NASA 2019b). Two of the five Lagrange points are defined as stable (L4 and L5), while the other three are unstable (L1, L2, L3). L4 and L5 have been identified as the most critical points the United States will most definitely need to control as they lead and follow the Earth's orbit (Ibid.). NASA believes L3 to be of minimal use at this time because it is always hidden behind the sun (Ibid.). On the other hand, L1 has an uninterrupted view of the sun, making it important for research. At the same time, L2 is ideal for astronomy because "a spacecraft is close enough to communicate with Earth readily" and because it has a clear view of deep space (Ibid.).

All of the areas identified above are of high importance and must be in astropolitical control by the United States if it wants to be the dominant space power continuously.

Astropolitics is the new geopolitics and astrostrategy is the new geostrategy. Americans must rid themselves of the notion that space is a vast, empty vacuum where the levers of international affairs do not operate. They operate in an amplified and different fashion, but they are no less real.

## Commit to Thwarting Foreign Threats by Adversarial Nations, Actors, Groups, and Criminals

In order to ensure space dominance, the United States will need to be committed to preventing threats from foreign nations, groups, or other actors in space. In a report by the White House named "United States Space Priorities Framework," released in December 2021, the administration shares a plan for its agenda for space. The report rightly states that the space capabilities of the United States enable the military to protect and defend the U.S. homeland, making it a crucial factor. Moreover, it advances "the national and collective security interests of the United States and its allies and partners" (The White House 2021).

As previously discussed, Russian and Chinese interventions in space are the most severe threats to the United States in space and on Earth. Therefore, the United States and its allies must protect themselves from possible attacks and interventions by these countries. According to the U.S. Space Priorities Framework, "The United States will enhance the security and resilience of space systems that provide or support US critical infrastructure from malicious activities and natural hazards." In order to maintain space superiority, the United States has also joined Australia, Canada, France, Germany, New Zealand, and the United Kingdom in the "Combined Space Operations (CSpO) Vision 2031." This initiative will partially focus on "the increasingly comprehensive and aggressive counterspace programs of other nation-states" and other threats presented by technological advances (U.S. Department of Defense 2022). However, state actors are not the only adversaries. It must be fully understood that the United States will face a myriad of non-state actors, terrorist groups, and criminal enterprises that Space Force will be required to neutralize. The environment that will decrease the cost of space activity and the lucrative nature of space will create the same conditions as they do on Earth for those who wish to commit acts of evil and violence.

## Build a USSF Fleet of Various Sizes and Compositions

Since the U.S. Space Force officially became the sixth branch of the U.S. military in December 2019, it has yet to figure out various issues but has started to build an agenda for the next 30 years. Twenty-three parts of the Air Force will relocate into the Space Force within the following months. The current fleet of only two robotic X-37B spacecraft and approximately 70 other spacecraft will expand significantly (Cohen 2020). X-37B has been described as an "orbital test vehicle designed to develop and demonstrate a wide range of long-duration space technologies" (Becker 2021). Unfortunately, the delta-v, a measurement to identify the ability of a spacecraft to maneuver in space, is not substantial, which is why future spacecraft will need to improve vastly (Becker 2021). According to Rep. Jim Cooper (D-Tenn.), the chairman of the House Armed Services Strategic Forces Subcommittee, the time to act is now, as he has stated that "the clock is ticking because the United States has been inviting an orbital Pearl Harbor for decades" (Becker 2021).

SpaceX has diligently worked on new orbital vehicles named *Starship* and *Superheavy*, both of which can be used in low Earth orbit for significantly less cost than any previously used spacecraft. These spacecraft will be considerably larger, ranging from 164 feet tall to approximately 400 feet when fully stacked (Becker 2021). Furthermore, these spacecraft can also operate in deep space and be fully reusable—unlike any rocket before. SpaceX will also build multiple versions, including a cargo version, which allows for manned spacecraft for orbital or deep-space missions, and a lunar lander. Within the next 30 years, SpaceX is planning on building 1,000 starships that the USSF will be able to add to its fleet (Becker 2021).

The most impressive addition to such new technology for the USSF might be a "massive increase in surplus maneuver," which will allow the Space Force safer, more flexible, and expansive operations in cislunar space and beyond (Becker 2021). The previously mentioned delta-v of such a starship designed by SpaceX is expected to be 6,500 meters per second, a drastic improvement compared to the range of an average satellite of 50–100 meters per second per year (Becker 2021).

In conclusion, representatives of the U.S. Congress have realized that the U.S. Space Force will need a fleet just like the U.S. Air Force or other branches of the military. As the space has previously only been used mainly for research purposes, the United States is very vulnerable due to a lack of strategic defense in the area. This must change very soon as other nations such as Russia and China have the military power to destroy satellites and other extraterrestrial resources the United States is currently using. Pat Bahn, the CEO of TGV Rockets Inc., stated, "The Space Force must be equipped with a fleet of responsive, spacefaring vehicles under the operational purview of the Space Force's equivalent of an Air Force colonel or Navy captain" (Bahn and Kyger 2020).

Further, a USSF Space Fleet requires a vision beyond current technology. Whether this is based on nuclear propulsion or propulsion that is now experimental or only conceptual, the USSF will ultimately need to possess a starship battle fleet to create the conditions for space power and to protect the NewSpace economy.

## Provide Law and Order for Civil and Commercial Enterprises

With the end of a binary existence in space, the geopolitical landscape of space has changed dramatically (Blount 2011). An expansion of space exploration into multiple states, specifically in Asia (e.g., Korea, Japan, and China), and up-and-coming commercialization have outdated law and order in extraterrestrial space and must, therefore, be adapted and reformed in a way that does not endanger some countries more than others (Blount 2011). Though the UN General Assembly has ratified the UN treaties mentioned previously, they are not fit for a "new space race" (United Nations 2021). Therefore, the United States must provide law and order for civil and commercial enterprises within this space (Blount 2011).

A new sphere that the United States will need to face is commercial actors with different interests than the state. NewSpace companies, such as Virgin Galactic, are trying to step outside of the usual governmental contracting and focus on space tourism and other enterprises (Blount 2011). This challenges the current state of law and order in space and the approach the United States will need to take when creating new policies, but it also makes an issue of ideas of security for the United States. A problem such as the definition of property rights in celestial bodies is only one of many that must be faced when providing new civil and commercial enterprises laws (Blount 2011).

Attempts by the United States, such as the Artemis Accords of 2020, have also not been welcomed internationally (Blount 2011). However, the Artemis Accords are a good touchstone for the United States to use, and if other powers don't wish to utilize this, it is their problem.

Charlie Bowles from EM Law stated, "The status and liability of commercial use of outer space, including the moon and other celestial bodies, is not very clear under the existing space law regimes" (Bowles 2021). For example, the current law described in the Liability Convention of 1972 deems the country in which the launch of a spacecraft occurs liable for all activities in outer space—even for nongovernmental activities (Bowles 2021). This will most likely be another aspect that must be changed with the commercial expansion to space (Bowles 2021).

The Space Force will need to act as both constabulary and magistrate in space. The physical distances in space and the potential problems of communication will require us to return to a time when diplomats and soldiers operated semiautonomously as they had to for millennia.

## Achieve Maximum Space Domain Awareness

Achieving maximum space domain awareness (SDA) is a critical issue for the United States within its long-term doctrine for space because it will ensure national safety and military prowess. With the first public accident between two artificial satellites occurring in 2009, this topic has been highlighted within the last decade and must be taken extremely seriously. In addition, China's firing of an SC-19 ASAT missile at its own weather satellite, creating over 15,000 pieces of debris in low Earth orbit, has brought further attention to SDA and the importance for the United States to have maximum awareness (Di Mare 2021).

The exponentially rising numbers of satellites in space (e.g., SpaceX 15,000 satellites, Blue Origin 4,000 satellites) will make space domain awareness an even more vital aspect of space policy within this century (Mann, Pultarova, and Howell 2022).

The space domain is separated into three capabilities that the United States will need to excel in if it wants to be the consistent space power. The first is space surveillance and tracking, including conjunction analysis, fragmentation, and reentry (Di Mare 2021). Second, space weather and the study of solar activities must be won by the United States, as it can heavily influence the performance of "space-borne and ground-based technological systems" (Di Mare 2021). The third capability is space intelligence, which means collecting data and information and conducting analysis on unknown satellites (Di Mare 2021). The latter of the three capabilities is specifically crucial for command and coordination within space.

According to the China Aerospace Studies Institute, space domain awareness can effectively defend potential actions from competing space powers in five key ways. According to the institute's report, SDA "enhances deterrence and reduces inadvertent escalation," allows for warfighting, can assist in protecting and developing norms and regulations pertaining to space activity, maintains space sustainability, and enhances the influence and role that the United States might play in space (Pollpeter 2021).

The Space Force identifies data as the biggest challenge for SDA. Collecting, synthesizing, fusing, and making sense of such an abundance of data is a challenging task that the USSF must master to maximize its SDA (U.S. Space Force 2020). Yet, despite this challenge, identifying different trends, patterns, and associations may provide the United States with information essential to succeed in space and defeat its adversaries in potential future space wars (U.S. Space Force 2020, 39).

### Create a Rapid Reaction Force, Space-trained Forces, and Space Intelligence, Including Covert Operations

As outlined in chapter 4, either a fully autonomous Space Force intelligence service or an entirely independent Space Intelligence Service will be required. During its first two years in existence, the U.S. Space Force has made some progress in space intelligence. In January 2021, the USSF was added to the U.S. Intelligence Committee, which comprises 18 members (Underwood 2021). Additionally, the USSF has created the National Space

Intelligence Center (NSIC), also known as the S-2 (Underwood 2021). According to Maj. Gen. Leah Lauderback, the director of Intelligence Surveillance and Reconnaissance (ISR), the NSIC "will perform national and military space foundational missions and evaluate capabilities, performance, limitations, and vulnerabilities of space and counter-space systems and services." The space intelligence gathered by the NSIC will be shared with senior leaders in the Department of Defense, a variety of policymakers, and warfighters of the USSF (Underwood 2021). The Space Intelligence Community of the United States includes multiple groups—the Defense Intelligence Agency, the National Security Agency, the National Geospatial-Intelligence Agency, the National Reconnaissance Office, and as of January 2022, the NSIC.

Furthermore, the USSF has started its intelligence work in space within the past year (2022). In January 2022, they launched a space-based space surveillance (SBSS) satellite, which will be in charge of tracking other objects in the GEO and LEO (Munoz 2022). Lastly, internship programs for space intelligence have also started at the USSF headquarters. After graduating its first two interns in 2020, forthcoming interns will be part of a two-year space intelligence program that will expand their knowledge and capabilities regarding ISR (Whiting 2020).

Contrary to the efforts in space intelligence, the USSF is clearly lacking an outlined plan for a rapid reaction force of any kind. Though China displayed its technological abilities to shoot down satellites more than a decade ago, it does not seem as though the USSF has the needed gear and inventory "to respond quickly to national crises" (Beames 2021). This is an anomaly, as all other military services have all the necessary resources (Beames 2021). According to Charles Beames, a former senior official in the Department of Defense, "A rapid response space capability must be an integral component of the president's response options to not only deter an adversary but also to avoid repeating the humiliation of revealed impotence in the face of a threat" (Beames 2021). He continues to say that the only way for the United States to defend itself against enemy attacks potentially will be through "quick reaction satellites and agile launchers" (Beames 2021). Those would allow the United States, even after being attacked, to "reconstitute at least a partial capability within a day" (Beames 2021). However, the current response time for such an attack would be approximately three years. Moreover, the commercial sector of the United States has been developing tools for an excessively faster response time for multiple years now. Still, the most recent budget plans of USSF do not include the needed funds to make such acquisitions (Beames 2021).

Another important aspect of space warfare will be creating space-trained special forces. In August 2021, the Air Force activated the final field of the USSF, the Space Training and Readiness Command (STARCOM) (Secretary of the Air Force Public Affairs 2021). According to the secretary of the

Air Force, this will be "preparing guardians to prevail in competition and conflict through innovative education, training, doctrine, and test" (Secretary of the Air Force Public Affairs 2021). The individuals in STARCOM will be prepared to be "combat-ready USSF forces to fight and win in a contested, degraded, and operationally-limited environment" (STARCOM n.d.).

Nations allied to the United States have also considered rapid reaction forces for outer space. Germany, the Netherlands, Portugal, Finland, and Slovenia (all members of the EU) also created an initiative for a rapid reaction force. According to a report, "the new force is expected to include space and cyber capabilities and special forces and air transport" (Carter 2021).

The need for a single unified intelligence service that possesses both covert action and special forces capability will be one of the only ways to ensure American space power.

### Protect Allied and Civilian Human Presence, Exploration, Research, and Colonization

With further aggressive behavior from states such as China and Russia, the United States will need to act as a policy enforcer in protecting civilian human presence, exploration, and research within the next century. The Committee on Space Research (COSPAR) keeps an updated list of guidelines to protect "future research and scientific investigation of celestial bodies" (research outreach 2020). By following these guidelines, the United States ensures that no space actor returns any materials from outer space that might endanger Earth. Furthermore, no actors shall risk any celestial bodies by bringing dangerous resources from Earth into outer space.

The protection of human presence will also play a more prominent role in this century. In 2021, companies such as SpaceX and Blue Origin launched their first all-civilian crews into orbit (research outreach 2020). The United States and its allies will need to focus on new policies to protect these civilians in outer space by granting them the freedom to travel within this space while not being in danger of any human-made satellites or other weapons in space. The Outer Space Treaty states that "Outer space, including the moon and other celestial bodies, shall be free for exploration and use by all States without discrimination of any kind, based on equality and following international law, and there shall be free access to all areas of celestial bodies." It will be imperative for the United States and its allies to keep this part of the treaty and protect such space missions. In addition, in its memorandum of understanding, NASA and the USSF have made it essential to highlight the importance of the "Search, rescue, and recovery operations for human spaceflight" (NASA 2020).

Furthermore, the exploration of space for reasons of research and the betterment of the human species will also be a vital factor that America and its allies will need to protect. The Planetary Protection Policy has been put in place by COSPAR to "protect the space environment from 'harmful contamination' which would endanger the integrity of the scientific exploration of outer space including the search for life" (Cheney et al. 2020). This policy differentiates space missions into low-risk and high-risk missions. Low-risk missions are defined as those with a "target which is not of direct interest for research into evolution or the origin of life" (research outreach 2020). On the other hand, high-risk missions require more protection to "ensure its protection from backward contamination from other planetary bodies" (research outreach 2020). High-risk missions also are essential to outer space research.

Within this century, the United States will need to emphasize protecting the liberties of space while dominating the area. As the United States has always stood for individual rights and liberties on Earth, the state will need to ensure the same in outer space as technologies advance. This will allow for more outstanding research and progress in the area.

Ultimately, as American and allied colonies are established, Space Force will need to act initially as the law enforcement and military protector of these colonies.

### Create Doctrine and Strategy beyond Cislunar

In September 2020, the Space Force and NASA signed a memorandum of understanding (MOU) that will highlight the collaboration on projects, allowing the United States to travel to cislunar space and beyond in a safer, more secure fashion. Per the document, "new US public and private sector operations extending into cislunar space, the reach of USSF's sphere of interest will extend to 272,000 miles and beyond—more than a tenfold increase in range 1,000-fold expansion in service volume" (NASA 2020). The two parties will continue their partnership in a variety of areas, but they will also cooperate in new fields proven to be essential for space dominance. Some of these areas include surveying and tracking deep-space technologies, which will develop further SDA and NEO detection (1), and "Interoperable spacecraft communications networks for Earth orbit and beyond" (2) (NASA 2020).

Additional to its efforts displayed in the MOU, NASA launched a giant antenna (DSS 53) in February 2022. This will allow engineers and other researchers on Earth to communicate "with the growing number of spacecraft exploring our solar system" (Jet Propulsion Laboratory 2022). The importance of the 30-foot antenna is enormous, considering its far reach into deep space. Scitechdail.com writes that "the DSN [Deep Space Network] allows missions to track, send commands to, and receive scientific data from faraway spacecraft" (Jet Propulsion Laboratory 2022). The 14 active antennas of the United States are expected to support approximately 80 operations that will launch within the next few years (Jet Propulsion Laboratory 2022). King Felipe VI of Madrid, Spain, attended the inaugural ceremony for the antenna in March (Jet Propulsion Laboratory 2022).

The USSF has also signed a contract with Northrop Grumman for new deep space radar capabilities. Focused on giving details on all-weather and constant coverage of all objects in GEO, it will provide essential information to the sixth member of the armed services (Jet Propulsion Laboratory 2022).

The Cislunar Highway Patrol (CHPS) is a satellite that will be launched into cislunar space. According to arstechnica.com, "this satellite is the beginning of an extension of operations by US Space Command from geostationary space to beyond the Moon" (Berger 2022).

These are all embryonic steps. It has been hard enough to move the space professional community from LEO/GEO to cislunar. This is while many non-space professionals are not even at this level. However, as many are now obsessed with cislunar, the time is now to think beyond it, beyond Mars, and conceptually beyond the solar system. We will either define the terms and the environment, or the terms and environment will be defined for us.

## MID-TERM DOCTRINE FOR THE COMING DECADES TO THE MIDDLE OF THE CENTURY

### Make a Case to the American Electorate

In chapter 1, I made the case that space professionals have mostly ignored the American electorate. This is inefficient, anti-American, and dangerous. Even before the Space Force was created, voices of dissent arose for multiple reasons. Many were concerned with additional bureaucracy in the Department of Defense and higher costs (Kelly 2018). A poll done in 2018 by YouGov, an international data and analytics firm, indicates a clear dislike for creating a Space Force. Only 29 percent of the over 19,000 individuals surveyed believed it a "good idea" (YouGov 2018). Even prominent individuals such as Brandon Weeden, a former space-operations officer with USAF, obstructed President Trump's plans (economist .com 2019). Weeden, according to *The Economist*, has opposing views mainly due to clashes with "geographical-area commanders" (The Economist 2019). In addition, he worried "that a dedicated command might encourage those fixated on 'future battles in space' . . . rather than the more pressing task of using space assets to help commanders wage war on Earth" (The Economist 2019). For all of the reasons listed above, the USSF will need to emphasize the importance of space and convey such information to the electorate in various ways. Surveillance, natural threats, and economic interests from space are vital and should be explained so that the United States may elevate its Space Force.

Surveillance within space will be one of the first steps essential to American dominance and must therefore be highlighted by USSF. However, though surveillance is critical, American efforts cannot stop there. Aggressive actors will indeed be unlikely to attack when knowing that they are observed, but if the observation can't follow with any counterattack, it loses its power (P. Garretson 2021). Therefore, according to Peter Garretson, columnist for aerospaceamerica.com, "we can expect that the Space Force will develop a patrol craft capable of various responses" (P. Garretson 2021). Explaining the importance of this area to the American electorate will also establish the USSF and create a more positive relationship.

Natural threats to the planet will be another issue that the electorate will confront with USSF. For example, there always is a minor threat of an asteroid or comet hitting U.S. land. The American people will most likely believe that the USSF is responsible for such events, so they need to be trained in this area before emphasizing this point to the American electorate (P. Garretson 2021). Though challenging, this might be one of the best ways to describe the necessity of the USSF to the entire population (P. Garretson 2021).

Lastly, economic interests will play a significant role in the importance of USSF. Resources on asteroids and the Moon will prove vital at some point in the future when resources on Earth become scarcer. Like the economic importance of Europe's findings hundreds of years ago, these resources play an essential role in future life on Earth and beyond. Therefore, the U.S. Space Force will need to protect them at all costs (P. Garretson 2021). This also provides USSF with an ideal opportunity to stress an area of space politics that will be essential for the United States to stay in power.

However, not involving the American electorate is also a titanic lost opportunity. Space Force can ignite the imagination of generations of current and yet-to-be-born Americans about a new frontier full of new opportunities, new hope, and the ever-expanding greatness of American civilization.

## Protect and Ensure the Resilience of Current Space Assets

At this point, over 36,000 objects orbiting the Earth are larger than tennis balls—of which only 13 percent are controlled (European Space Agency 2022). Though this might not seem like a significant issue, they bear a great threat to satellites used by the United States and other nations—many of which our economy and, more importantly, our society depend on regularly (European Space Agency 2022). Therefore, the United States must be able to protect and ensure the resilience of current space assets as soon as possible.

Some of the most critical satellites (valued at more than $1 billion) are highly vulnerable at this time, creating an extreme risk for the United States and its allies (Starling et al. 2021). One strategy the United States has considered is deterrence-by-denial (Starling et al. 2021). According to the Center for Space Policy and Strategy, it "attempts to deter undesired behavior by leading actors to conclude that they will be unable to achieve the objectives they seek from their behavior" (Gleason 2021). By investing in technology that makes attacks on space undesirable, the vulnerability of such space assets could quickly be mitigated, giving the United States a clear upper hand.

Another strategy could be deterrence-by-punishment. With this approach, "actors may be deterred from undesired behavior if they conclude that the costs of the behavior outweigh the benefits" (Gleason 2021). A straightforward example of such an approach would be NATO and its official documents (Noll, Bojang, and Sebastiaan 2020). Pertaining to space, a message in the U.S. National Security Strategy indicates a straightforward approach: "Any harmful interference with or an attack upon critical components of our space architecture that directly affects this vital US interest will be met with a deliberate response at a time, place, manner, and domain of our choosing" (Gleason 2021).

A final issue the United States has been facing—unrelated to aggressive actors in space—is the amount of debris in orbit. Whipple shields are methods that the ESA (European Space Agency) has used before. They function as bumper shields that "cause the impactors to completely disintegrate during impact" (European Space Agency n.d.).

The importance of securing and protecting space assets from a different sphere—namely cyberattacks—will also prove to be vital for the United States to maintain its rule as the world hegemon. The formation of a Space Systems Critical Infrastructure Working Group, a mixture of government and industry leaders in the space, is a good beginning for such protection (Brooks 2022). When developing the group, the CISA stated, "the working group will serve as an important mechanism to improve the security and resilience of commercial space systems" by finding solutions and mitigating risk to space assets (Brooks 2022).

Finally, there must be doctrinal changes. An attack on American assets in space is an attack on the United States. Therefore, we must reserve the right to defend and attack any way we deem necessary to deter, disarm, or destroy any potential threat and view American space as no different from an American embassy or the American homeland.

### Support for Terrestrial-based Military and National Security Needs

In the coming years, the U.S. Space Force will need to ensure support for operations in extraterrestrial space and show support for terrestrial-based military and national security needs. According to Paul Szymanski from airuniversity.edu, issues have been abundant with the "integration of space and terrestrial military actions" in the past (Szymanski 2020). While terrestrial commanders have ranked space actions as low-priority compared to other actions with urgent requests, space-related warfighters "do not consider political consequences" (Szymanski 2020). An unhealthy back-and-forth between two military areas is not sustainable and must be resolved immediately.

Weaponry such as terrestrial sensors has also gained importance—an area that the USSF will need to invest in. According to the heritage foundation, the United States' number of sensors is insufficient considering the aggressive actors in space (Venable 2021). Though two more SSA satellites will be launched in 2022, the United States will need to emphasize increasing this number within the next years to support military and national security needs (Venable 2021).

There must also be a transition for defensive and offensive platforms in space that support the terrestrial fight to offset and shift any balance of power on Earth.

### Forge Critical, Dynamic Partnerships with NewSpace and the Private Sector

With the commercial sector of space picking up in the past years, it will be vital for the U.S. Space Force to maintain a good relationship with companies that provide a possibly valuable product or service. Col. Eric Felt, head of the Air Force Research Laboratory's Space Vehicles Directorate, believes that USSF will follow the NASA route of collaborating with the private sector while creating its own technology (Erwin 2020b). The past two years have proven that USSF is more than willing to work with companies such as SpaceX, Rocket Lab, Blue Origin, or ULA—a promising sign for the future (Alamalhodai 2021).

USSF has relied on the private sector significantly in the area in which data analytics systems and sensors track space objects (Erwin 2020a). Numerica, a space data provider that "operates a network of deep space telescopes," is only one example of dozens of companies aiming its services at government services (Erwin 2020a). Low Earth orbit satellite tracking and daytime tracking are areas that Numerica has moved into to make it more attractive for government contracts (Erwin 2020b). This seemingly warm approach from both sides will help the United States build long-lasting, dynamic partnerships with many private companies, pushing efficiency in prices and operations while helping the economy.

One of the most significant areas for government contracts is space domain awareness (SDA). Col. Felt emphasizes the robust sensors and data analytics systems many private companies have in place, which can help study space objects (Erwin 2020b). Companies such as Space Domain Awareness, Savid, Astroscale, Astra, and Digantara are all working in this field to provide the Space Force with valuable information and help the United States stay the supreme leader in space (Craft n.d.).

An interesting idea that has been discussed is the possibility of a "space commodities exchange" (Erwin 2020b). In this exchange, space services provided by private companies would be traded just like commodities are traded on the Chicago Mercantile Exchange. Col. Felt advocates for this, saying, "It opens up the financial engine to optimize the price and the quality, where you establish certain quality standards for what you're going to need. So . . . the space domain awareness data might be a great example of the kinds of things that the Space Force could purchase through a space commodities exchange" (Erwin 2020b).

### War-game and Simulate Future Crises and Conflicts in Space with Current and Future Capabilities

The significant amount of war games it conducted helped the U.S. navy during the period between World Wars I and II. With approximately 300 war games, Admiral Chester Nimitz once said after the victorious ends of World War II, war-gaming played an enormous role. It guaranteed that "nothing that happened during the war was a surprise . . . except the kamikaze tactics" (Bilsborough 2020). A similar approach must be taken for space war games, as the space race is heating up and other state actors are working on their tactics to destroy American space assets.

Since 2001, the United States has held the Schriever Wargames, an annual event that challenges leaders in the field of space with realistic war games. In 2019, for example, the Schriever Wargame was set in 2029 with five objectives the participants must complete. With over 350 attendants from 27 worldwide agencies, the Schriever games play a significant role in the field of war games for space (Hill 2019). Projects like this and others must become standard practice for the USSF to be aware of all potential scenarios fully.

In December 2021, the most recent war game tested "satellite resiliency," according to Reuters (Stone 2021). Some of the issues worked on were the "shooting down of US missile-tracking satellites, satellite jamming, and other warfare tactics that could possibly be used in space" (Stone 2021). Thirteen similar exercises have taken place so far with allied nations, including Australia, Britain, and Canada (Stone 2021).

Shane Bilsborough from warontherocks.com has identified six different objectives the United States should be organizing these events around

(Bilsborough 2020). By following these, the United States will gain substantial competence that will allow it to stay in charge of space for many years to come. The six objectives are:

1. Understanding the geography of space, and the power to describe it
2. Assessing how space assets could possibly operate in teams
3. Exploring the role of commercial space systems
4. Anticipating technology competitions
5. Tie space operations to combat on Earth
6. Surfacing new concepts, theories of victory, and redlines (Bilsborough 2020)

These space war games need to focus on the present, the midterm, and the long term. The midterm and long-term war games need to utilize forecasting, prediction, and futurism to game out potential war games with weapons and concepts that do not exist yet. We must break out of the curse of presentism, so infectious throughout American society and even within aspects of the military.

### Active Steps and Commitment to Achieving Space Dominance and Governance

In order to achieve one of the goals—space dominance—the U.S. Space Force will need to follow a very detailed plan that includes short-, middle-, and long-term strategy, including milestones. The Atlantic Council has published a paper on U.S. strategy for space until 2050. It discusses many vital milestones the United States will need to achieve to gain dominance in space and secure liberty for all actors involved (Starling et al. 2021).

While acknowledging the stark drift of the politics of space from exploration to security and commerce, the Atlantic Council states, "Security in space is at risk, and the United States must act urgently" (Starling et al. 2021). Some of the most important steps the United States will need to commit to achieve space dominance are creating laws for safer operations in space, defending space access from nations that would deny it (e.g., China, Russia), focusing on regulation and targeted investments that will quicken space commerce, and gaining complete power over cislunar space promptly (Starling et al. 2021).

The U.S. Space Force has also released a Campaign Support Plan (2021), written by the Air Force Public Affairs Secretary. Its primary efforts are to "support Geographic Combatant Commands (GCCs) through various ways to prepare Space Force" (U.S. Space Force 2021). Moreover, the plan takes a "three-pronged approach" with the main efforts being to expand relationships of allies, strengthen allied partnerships to "increase cooperation, collaboration, and interoperability in the space domain," and to "leverage relationships with competent partners in operations and acquisitions" (U.S. Space Force 2021).

### New Treaties, Laws, and Language—Farewell to the Outmoded Outer Space Treaty

Within the last 55 years, space exploration and its politics have changed substantially. Individuals have landed on the Moon, launched hundreds of satellites, and commercial actors have joined the new space race. Many of today's outer space operations could not have been foreseen in such a way in 1967, which is why many individuals are concerned with the laws established in some of the outmoded treaties signed during the Cold War—the Outer Space Treaty being one of them.

Since then, the United States has made an effort to reform outmoded systems of this kind by creating newer methods such as the Artemis Accords, signed in October 2020. Though not accepted by all space actors, specifically China and Russia, the Artemis Accords are a new set of governing standards that the United States would like to impose in space to make it safer for all actors involved. Since its inception, states' reactions to the accords have been mixed (Deplano 2021). An article written by Rossana Deplano argues that "while rooted in the provision of the Outer Space Treaty, the Artemis Accords introduce a significant innovation in international space law by replacing the anticipatory approach to the regulation of outer space activities with the staging principle of adaptive governance" (Deplano 2021). Although the accords contain commitment to essential issues such as astronaut rescue programs, the United States and its allies will need to craft new treaties ratified by the United Nations to hold the entire world accountable.

Even the United Nations believes it to be time for a new, updated version of important space documents such as the Outer Space Treaty (Hanlon and Autry 2022). Explaining that peace in space "remains uncomfortably tenuous," Michelle L. D. Hanlon and Greg Autry stress the importance of new laws in an article on cnn.com (Hanlon and Autry 2022). One of the more forceful critiques the two authors have is that the Outer Space Treaty does not "offer detailed 'rules of the road'" (Hanlon and Autry 2022). In 2021, the United Nations started five draft resolutions about the nonproliferation of space and peaceful use of action. Another issue discussed is the use of chemical weapons in outer space (United Nations 2021). The approval of draft resolutions by a landslide majority of 163 to 8 is a promising sign that treaties for space governance might be signed into action within the next few years (United Nations 2021).

Another research project developing plans for space law and military space operations is the Woomera Manual (Woomera 2018). Universities from the United Kingdom, the United States, and Australia—namely the University of Adelaide, the University of Exeter, the University of

Nebraska, and the University of New South Wales-Canberra—have come together with the mission to "develop a manual that objectively articulates and clarifies existing international law applicable to military space operations" (Woomera 2018).

In the end, any American adherence to international law must be based on placing American and allied interests and values front and center. Treaty regimes that tie American and allied hands while allowing the Axis of Evil, Axis of Resistance, or the like is a non-starter.

### Dominance of LEO, GEO, and xGEO

Dominance over the three regions of low Earth orbit, geosynchronous orbit, and beyond geosynchronous orbit must be approached from two perspectives. First, and most important, the United States will need to dominate all three of these areas militarily to create a political space climate in which all can profit. Second, the United States must take advantage of the economic aspect of space to grow its economy and stay dominant on Earth.

According to a group of researchers from spaceforcejournal.org, "Strategists and policymakers must account for US interests beyond geostationary orbit (XGEO) when developing future concepts of operations, policies, and resourcing for space forces" (Buehler et al. 2021). A second gravitational body in xGEO will be a physical challenge to overcome, but the USSF must prepare to defend this area in space just as LEO and GEO (Buehler et al. 2021). With over 80 percent of the space economy being commercial, the USSF will have to work with industry leaders to establish the best space assets possible and, therefore, control this area (Buehler et al. 2021). Concerning the economy, xGEO will prove to be vital in space tourism and asteroid mining—the latter of the two will improve the economy and provide the United States with new resources on Earth (Buehler et al. 2021).

Sometimes referred to as the "most valuable real estate" in space, satellites in this area can stay at a fixed point in relation to Earth and can therefore observe the same region at all times (Buehler et al. 2021). Consequently, it is essential that this area be controlled by the USSF if the United States seriously wants to dominate space. On March 1, 2022, the United States continued its attempts to dominate this space by launching another satellite, GOES-T/GOES-18, into GEO. It was successfully placed in GEO, 22,236 miles above Earth, on March 14, 2022 (NESDIS 2022).

Low Earth orbits most likely are the most competitive of the three areas. It is also where the United States is in most danger of losing dominance over (Odell et al. 2021). With an exponential increase in access to space within the past few decades, the immense number of competitors has increased and endangered the U.S. monopoly. In a paper from the Institute for Defense Analyses, Laura Odell and her colleagues suggest that the "DoD will need to work harder to maintain and secure national assets targeted by peer competitors engaged in gray zone competition" (Odell et al. 2021). China is the biggest threat to the United States in this regard. Defensenews.com believes that the ideal way for the United States to maintain its dominance is by investing "sufficient resources in preparing its new Space Force" (Zivitski 2020).

**Establish the Lunar Base and Expand the Lunar Presence**

NASA's Artemis program plans to land humans on the Moon again by 2024. After successfully completing Artemis 3 within the following years, NASA wants to build a base on the surface of the Moon—a possible stop for future missions to Mars and other planets (Royal Museums Greenwich n.d.). This lunar base camp is supposed to house American astronauts for as long as one to two months at a time and equip them with rovers that could help them with lunar research (U.S. Government Accountability Office 2021). Most likely located on the lunar South Pole, the base would provide astronauts of the United States with rovers and a habitable place. Still, it will also establish the United States as an economic force to be reckoned with in space (NASA 2019a). According to NASA, the Artemis Base Camp that the United States will try to develop will need to face Earth to simplify radio waves and other forms of communication for engineers (NASA 2021). According to NASA's Plan for Sustained Lunar Exploration and Development, "the base camp will demonstrate America's continued leadership in space and prepare us to undertake humanity's first mission to Mars" (NASA 2019a).

The Artemis Base Camp will have three main elements. The lunar terrain vehicle (LTV) that will be able to move around the crew (1), a habitable mobility platform for longer trips away from the base camp (2), and "the foundation surface habitat" (3) are all key elements that will allow the United States to research the lunar south pole in more detail.

The United States cannot become complacent, though. China has publicized plans to create a lunar research station by approximately 2027, which should concern the Western world considerably (Chen 2021). Though previously scheduled for 2035, the Chinese regime has emphasized space and is trying to destroy U.S. space dominance by all means. By collaborating with Russia, China is working on developing the International Lunar Research Station (ILRS) (Pultarova 2021). The station will supposedly exist as a space station in lunar orbit, a moon base, and mobile rovers (Pultarova 2021).

A lunar base will ultimately lead to the first American colony in space, and this must be the place to expand American and allied presence on the Moon and create the military, legal, economic, and scientific governance of the Moon by America and allied powers that can be used as a model for Mars and beyond.

### Dominate the Lagrange Points

As discussed previously, the dominance of USSF over Lagrange points will be imperative to success in space. More specifically, L4 and L5 will play the most pivotal role for USSF as these are the only two "stable" points (Howell 2017). The European Space Agency (ESA) describes the two points as "a ball in a large bowl" where asteroids and space dust tend to amass (Howell 2017). The fourth and fifth Lagrange points, according to Gerard O'Neill, could potentially provide humans with the area for a new space colony (Howell 2017).

The importance of Lagrange points for state actors lies in the ability to launch spacecraft or satellites into specific areas. They will be able to stay in one stable orbit while being in a highly fuel-efficient space (Kaplan 2020).

China has successfully launched its Queqiao relay satellite into the Lagrange point, trailing the Moon (Jaeger 2019). With this, the U.S. Space Force must create a step-by-step plan to calculate how to dominate these LaGrange points, ensuring space dominance.

### Forge Partnerships with NewSpace to Begin to Build the Critical Infrastructure

The military must understand and partner with NewSpace as much as NewSpace needs to connect with astropolitical situations and dangers that only the Space Force can solve. The launch of rockets and their technology is significant, but an aspect of space that has not received nearly as much attention is infrastructure. In the private sector, dozens of companies have specialized in various areas that will be deemed necessary in the future, when space has advanced even more. By 2030, most American banks project a minimum revenue of $1.4 trillion of revenue in the space industry (Rossettini 2021). With an abundance of undiscovered issues, the United States will need to ensure good partnerships with the private sector and create contracts for companies that build critical infrastructure in space. From projects that would enable commercial internet networks in space to companies that focus on traffic management, USSF must not overlook this vital aspect of space (Tonar and Talton 2018).

### Begin to Build Reusable Spacecraft and "Starships" and Engage in the First Patrols

After the success of Space Shuttle in the years from 1981 to 2011, the USSF must find new success in reusable spacecraft (NASA n.d.). SpaceX has worked on a reusable rocket, Falcon 9, which will be able to launch humans into space as well. According to SpaceX CEO Elon Musk, some of the pieces on Falcon 9 should even be reused as often as 100 times (Musk 2016). According to SpaceX's website, as of March 26, 2022, Falcon 9 has launched a total of 144 and landed 104 times (SpaceX 2022). Another partially reusable spacecraft is the X-37B Orbital Test Vehicle (Secretary of the Air Force Public Affairs 2020). Used as a test program by the U.S. Air Force, its main objectives are to further reusable spacecraft technology and to operate experiments that "can be returned to, and examined, on Earth" (U.S. Air Force n.d.).

China had entered the field of reusable spacecraft as well in 2020 when it launched "its own Boeing-like reusable space plane" that may have had "as many as seven crewed and non-crewed spaceplane projects in development" (Piesing 2021). In addition, according to Space.com, China wants its new rockets for astronauts to be reusable (Jones 2022). Such new technology for China "would allow a reusable launch option for sending astronauts or cargo to China's new Tiangong space station. In addition, a larger version would allow China to send a crew on lunar and deep space missions" (Jones 2022).

The U.S. Space Force has made significant progress in the technology of patrolling as well. For example, the new cislunar Highway Patrol System (CHPS) previously mentioned will be used to patrol around the Moon, therefore being some of the first patrols of the USSF. The three main functions of the CHPS will be "detecting, tracking, and identifying manmade objects in this vast span" (Seck 2022).

### Begin to Build a Space Intelligence/Special Forces Cadre

When 2nd Lt. Samuel Pisney started attending USAF's intelligence officer course in September 2020, it was an essential step for the USSF (Rieves 2020). The first-ever U.S. Space Force intelligence officer will be trained in intelligence, surveillance, and reconnaissance (ISR) and will begin a new chapter in the Intelligence Department of the USSF (Rieves 2020).

When discussing ISR at an Air Force symposium in 2021, Gina Ortiz Jones, the undersecretary of the Air Force, highlighted the importance of intelligence and USSF (Hadley 2021). "When we think about ISR, I think what's important is ensuring that we are thinking . . . not just from the Air Force perspective, but also from the Space Force perspective" (Hadley 2021). As a result, efforts in this field have been vigorous. The USSF has already held ISR-focused conferences and other discussions focused on space intelligence and how USSF can succeed in this area (Lee 2020).

In the coming years, the Space Force will need to focus on graduating more individuals from intelligence officer programs and creating a group of highly knowledgeable people in the field of space intelligence so that

USSF can be protected from all scopes. USSF has taken the proper steps to build space intelligence by creating a unit specified for space intelligence. Delta-7 "is the operational Intelligence, Surveillance, and Reconnaissance (ISR) element of the US Space Force" and will play a vital role in future endeavors in the astropolitical field (Peterson Space Force Base n.d.).

## Build and Deploy a Complete Space-based Missile Defense System

The idea of a completely space-based missile defense system such as the Strategic Defense Initiative (SDI) President Reagan proposed was, for a long time, nothing but an idea laughed at and not taken seriously by most Americans. President Reagan might have been ahead of his time with the concept of a defense system against nuclear attacks solely controlled from outer space (Editors of Encyclopedia Britannica n.d.). With the inception of the U.S. Space Force, the idea of a totally space-based missile defense system has found ground again and should be in the American repertoire of national defense in the future.

Within the past two years, USSF has launched missile-warning satellites into orbit. The Space-Based Infrared Systems Geosynchronous Earth Orbit satellite (SBIRS GEO-6), a satellite built by Lockheed Martin, completed production in September 2021. Singh Bisht from thedefensepost.com does a phenomenal job of explaining the missile warning system: "The SBIRS is an infrared-based early warning satellite system that includes a combination of satellites and payloads in Geosynchronous Earth Orbit and Highly Elliptical Orbit. . . . The sensors will provide vast amounts of data to 'detect missile launches, support ballistic missile defense, expand technical intelligence gathering and bolster situational awareness on the battlefield'" (Bisht 2021).

The USSF is already working on the successor of the SBIRS. The "Next-Generation Overhead Persistent Infrared (OPIR) warning satellite" will be able to "utilize more capable sensors to detect enemy missiles through their heat signatures" (Bisht 2021). Overall, it seems as though the USSF is making significant progress in space-based missile defense systems, allowing for an utterly space-based system in the future.

All ideas from the SDI era need to be resurrected and acted on. In particular, action must be taken regarding space-based interceptors, brilliant pebbles, and the direct energy weapons (DEW) programs.

## Build a Robust System of Planetary Defense

With NASA being under congressional direction to "find and track 90 percent of NEOs 1 kilometer or more in diameter within ten years," since 1998, and another congressional direction since 2005 to find "90 percent of those 140 meters or more in diameter within 15 years," the agency had a lot of work in planetary defense it could not keep up with (M. Smith

2020). Therefore, by only finding approximately 40 percent of the estimated population instead of 90 percent, it is more than helpful that they have paired with the U.S. Space Force to work on planetary protection together (M. Smith 2020).

Recommendations discussed in conferences such as the International Academy of Astronautics biannual planetary defense conference should be taken seriously and considered as possible steps of action. Though the odds of an impact on Earth are minimal, NASA and USSF must identify all potential dangers, as the impacts on Earth can be severe (Mainzer 2017). According to Amy Mainzer: "Approximately 70% of NEOs larger than 140 m (large enough to cause severe regional damage) remain undiscovered. With the existing surveys, it will take decades to identify the rest. Progress can be accelerated by undertaking new surveys with improved sensitivity" (Mainzer 2017).

By investing in advanced search systems, Mainzer argues, the chances of impacts can be minimized even more (Mainzer 2017). Therefore, the USSF should build a robust planetary defense system by following the outline discussed by Mainzer.

However, a transition must be made to make this a purely Space Force mission. Defending planet Earth in all arenas is a military function, and ultimately Space Force will possess the necessary offensive and defensive weapons for doing so.

### Prepare, Plan, and Deploy for Search and Rescue

During a panel on March 4 at the AFA Warfare Symposium, USSF chief technology officer Lisa Costa discussed some of the most critical space-security activities currently being worked on. One of the six mentioned was "enhancing current services, including search and rescue; space commerce; and intelligence, surveillance, and reconnaissance" (Adde 2022).

In 2020, a memorandum of agreement (MOA) about satellite-aided search and rescue systems was signed by numerous agencies—namely NASA, NOAA, DAF (including USAF and USSF), and the USCG. According to the document, USSF is responsible for "assisting in planning, demonstrating and evaluating the effectiveness of MEOSAR" (SARSAT 2020).

Before a potential tragedy blindsides everyone, there need to be concrete plans to obtain the necessary capabilities to engage this function.

### Establish a Commensurate Budget for Current, Midterm, and Long-term Obligations

The essential part of the U.S. Space Force will be that it has a budget big enough to allow it to achieve the goals discussed. While still very young, the USSF has already asked for a hike in funding. For 2022, USSF requested $17.4 billion, which would equate to a 13.1 percent increase in funding year-over-year (Tadjdeh 2022). According to General John Raymond, three significant areas that need the financially backed improvements are "modernizing the force, readiness and continuing to build-out the service" (Tadjdeh 2022). Raymond argued: "We cannot get left in the starting blocks. We've got to move at speed, and I would encourage Congress to pass an appropriations bill so we can get out of the starting blocks and keep running" (Tadjdeh 2022).

These are ambitious beginnings for the USSF, but a large budget must be established for all of the obligations within the following years to be successful. The current defense budget proposal for 2022 included the entire amount requested by the USSF, amounting to 2.5 percent of the entire U.S. defense budget. According to a press statement by the Pentagon, "Competitors like China and Russia are challenging America's advantage in space by aggressively developing offensive weapons to deny or destroy US space capabilities in conflict" (Erwin 2021). It is hoped that the Biden administration understands the importance of USSF and is willing to work with the branch to ensure space dominance by the United States for the long-term future.

However, the budget is the currency of seriousness in Washington. The USSF budget will need to be exponentially expanded to achieve the mid-term and long-term goals outlined in this book. If America is serious about its national security, it will be serious about space power as the ultimate arbiter of this security, and the budgets need to reflect this.

## CONCLUSIONS

Everything is determined by the inevitable choice versus the vagaries of political expediency. The United States does not want other nations to make a choice. Our adversaries will pursue all of these related to dominance and militarization. If we do not make these choices, they will make our choices for us, but on their terms. What we do now will determine the destiny of the nation for centuries.

# Bibliography

114th Congress. (2015). *U.S. Commercial Space Launch Competitiveness Act.* https://www.congress.gov/bill/114th-congress/house-bill/2262

Adams, John (1765). "[February 1765]." *Founders Online*, National Archives, https://founders.archives.gov/documents/Adams/01-01-02-0009-0002 [Original source: *The Adams Papers*, Diary and Autobiography of John Adams, vol. 1, *1755–1770*, ed. L. H. Butterfield. Cambridge, MA: Harvard University Press, 1961, pp. 255–258.]

Adde, N. (2022, March 8). "Six Space Technologies the USSF Needs in Order to Maintain the US Advantage." Air Force Magazine. https://www.airforcemag.com/six-space-technologies-the-ussf-needs-to-maintain-the-us-advantage/

Alamalhodaei, Aria. (2021, September 27). "US Space Force Awards $87.5M to Rocket Lab, SpaceX, Blue Origin, ULA for Next-Gen Rocket Testing." Tech Crunch. https://techcrunch.com/2021/09/27/u-s-space-force-awards-87-5m-to-rocket-lab-spacex-blue-origin-ula-for-next-gen-rocket-testing/

Arms Control Association. (2017). "Arms Control and Proliferation Profile: China." https://www.armscontrol.org/factsheets/chinaprofile

Arnold, D. C. (2009). "Space and Intelligence." In Damon Coletta and Frances T. Pilch, *Space and Defense Policy*. New York: Routledge.

Asterank. (n.d.). https://www.asterank.com/

Axe, D. (2020, August 19). "China Is Probing Japan's Defense over the Disputed Senkaku Islands." Forbes. https://www.forbes.com/sites/davidaxe/2020/08/19/china-is-probing-japans-aerial-defenses-over-the-disputed-senkaku-islands/?sh=3itle8ad7f807dab

Bahn, P., & Kyger, T. (2020, January 22). "A Space Force Needs Spaceships." SpaceNews. https://spacenews.com/op-ed-a-space-force-needs-spaceships/

Bartels, M. (2018, August 29). "India Will Launch Its Own Astronauts to Space by 2022, Government Says." Space.com. https://www.Space.com/41657-india-will-launch-astronauts-in-2022.html

Beames, C. (2021, November 9). "An Easy Fix to Space Force's Most Glaring Vulnerability." Breaking Defense. https://breakingdefense.com/2021/11/an-easy-fix-to-space-forces-most-glaring-vulnerability/

Becker, J. (2021, May 19). "A Starcruiser for Space Force: Thinking through the Imminent Transformation of Spacepower. War on the Rocks." https://warontherocks.com/2021/05/a-starcruiser-for-space-force-thinking-through-the-imminent-transformation-of-spacepower/

Béraud-Sudreau, L. (2020, February 14). "Global Defence Spending: The United States Widens the Gap." International Institute for Strategies Studies. https://www.iiss.org/blogs/military-balance/2020/02/global-defence-spending

Berger, E. (2022, March 3). "The US Space Force Plans to Start Patrolling the Area around the Moon." Ars Technica. https://arstechnica.com/science/2022/03/the-us-space-force-plans-to-extend-its-operations-to-the-moon/

Beuker, I. (2018, February 23). "The Space Economy: Modern-Day Gold Rush 100x Bigger than Bitcoin." Igor Beuker's blog. https://igorbeuker.com/marketing-innovation-blog/the-space-economy-a-modern-day-gold-rush-100x-bigger-than-bitcoin-2/

Bilsborough, S. (2020, November 17). "More Space Wargames, Please." War on the Rocks. https://warontherocks.com/2020/11/more-space-wargames-please/

Bisht, I. S. (2021, October 6). "US Space Force Missile Warning System Ready for Launch." The Defense Post. https://www.thedefensepost.com/2021/10/06/us-space-force-missile-warning-ready/

Blank, Stephen J. (2011). *Russian Military Politics and Russia's 2010 Defense Doctrine*. U.S. Army War College Press. https://press.armywarcollege.edu/monographs/584

Blount, P. (2011, January). "Renovating Space: The Future of International Space Law." *Denver Journal of International Law & Policy* 40(1): 515–532. https://digitalcommons.du.edu/cgi/viewcontent.cgi?article=1160&context=djilp

Bowen, B. (2020). *War in Space*. Edinburgh University Press.

Bowles, C. (2021, March 8). "Space Law: The Commercial Space Race Begins." EM Law. https://emlaw.co.uk/space-law-the-commercial-space-race-begins/

Brands, H. (2015, August 26). "American Grand Strategy: Lesson from the Cold War." Foreign Policy Research Institute. https://www.fpri.org/article/2015/08/american-grand-strategy-lessons-from-the-cold-war/

Broad, W. J. (2021, January 24). "How Space Became the Next 'Great Power' Contest between the U.S and China." New York Times. https://www.nytimes.com/2021/01/24/us/politics/trump-biden-pentagon-space-missiles-satellite.html

Brooks, A., & Agrawal, R. (2021, December 31). "Becoming the Department of the Air and Space Forces." Space Force Journal. https://spaceforcejournal.org/becoming-the-department-of-the-air-and-space-forces/

Brooks, C. (2022, February 27). "The Urgency to Cyber-Secure Space Assets." Forbes. https://www.forbes.com/sites/chuckbrooks/2022/02/27/the-urgency-to-cyber-secure-space-assets/?sh=2339b43d51b1

Brown, C. Q. (2020, August). *Accelerate Change or Lose*. U.S. Air Force. https://www.af.mil/Portals/1/documents/csaf/CSAF_22/CSAF_22_Strategic_Approach_Accelerate_Change_or_Lose_31_Aug_2020.pdf

Buehler, D., Felt, E., Finley, C., Garretson, P., Stearns, J., & Williams, A. (2021, January 31). "Posturing Space Forces for Operations beyond GEO." Space Force Journal. https://spaceforcejournal.org/posturing-space-forces-for-operations-beyond-geo/

Cahan, B., & Sadat, M. (2021, January 6). *US Space Policies for the New Space Age: Competing on the Final Economic Frontier.* Politico. https://www.politico.com/f/?id=00000177-9349-d713-a777-d7cfce4b0000

Cammarata, S. (2020, June 6). "Russia and China Should Be Viewed as 'one alliance' in the Arctic, U.K. Defense Official Warns." Politico. https://www.politico.eu/article/russia-and-china-should-be-viewed-as-one-alliance-in-the-arctic-uk-defense-official-warns/

Campbell, C. (2019, July 17). "From Satellites to the Moon and Mars, China Is Quickly Becoming a Space Superpower." TIME. https://time.com/5623537/china-space/

Carlson, J. (2020). *Spacepower Ascendant: Space Development Theory and a New Space Strategy.* Independent publisher.

Carlson, J. (2021, February 1). Interview. (D. L. Colucci, Interviewer).

Carter, L. (2021, October 21). "Germany, 4 EU States Launch Military Reaction Force Initiative—Report." DW. https://www.dw.com/en/germany-4-eu-states-launch-military-reaction-force-initiative-report/a-59574641

Center for Strategic & International Studies. (n.d.). "Of the Russian Military Industrial Complex—Russian Roulette Episode #33." Podcast. https://www.csis.org/podcasts/russian-roulette/russian-military-industrial-complex-%E2%80%93-russian-roulette-episode-33-0

Center for Strategic & International Studies China Power. (2020a). "How Does China's First Aircraft Carrier Stack Up?" https://chinapower.csis.org/aircraft-carrier/

Center for Strategic & International Studies China Power. (2020b). "How Is China Modernizing Its Nuclear Forces." https://chinapower.csis.org/china-nuclear-weapons/

Chandler, D. L. (2011, October 26). "Shining Brightly." MIT News. https://news.mit.edu/2011/energy-scale-part3-1026

Chen, S. (2021, December 29). "China Speeds Up Moon Base Plan in Space Race Against the US." South China Morning Post. https://www.scmp.com/news/china/science/article/3161324/china-speeds-moon-base-plan-space-race-against-us

Cheney, T., Newman, C., Olsson-Francis, K., Steele, S., Pearson, V., & Lee, S. (2020, November 13). "Planetary Protection in the New Space Era: Science and Governance." Frontiers in Astronomy and Space Sciences. https://www.frontiersin.org/articles/10.3389/fspas.2020.589817/full

Cohen, R. S. (2020, October 7). "Space Force to Lay Long-Term Groundwork in Second Year." Air Force Magazine. https://www.airforcemag.com/space-force-to-lay-long-term-groundwork-in-second-year/

Colarossi, N. (2020, March 4). "Temples, Opera, and Braids: Photos Reveal What China Looked Like before the Cultural Revolution." Business Insider. https://www.businessinsider.com/photos-china-before-communism-cultural-revolution

Collins, J. (1973). *Grand Strategy: Principles and Practices.* Annapolis: Naval Institute Press.

Colucci, D. L. (2018, August 20). "American Doctrine: The Foundation of Grand Strategy." *World Affairs.* Sage Journals. https://journals.sagepub.com/doi/10.1177/0043820018790793

Commission to Assess United States National Security Space Management and Organization. (2001, January 11). *Report of the Commission to Assess United States National Security Space Management and Organization.* https://aerospace.csis.org/wp-content/uploads/2018/09/RumsfeldCommission.pdf

Cooper, A. H. (2021, January 19). Interview. (D. L. Colucci, Interviewer).

Cox, M., & Stokes, D. (2008). *US Foreign Policy.* Oxford University Press.

Craft. (n.d.). "Space Domain Awareness." https://craft.co/space-domain-awareness

CSIS Missile Defense Project. (2021). "RSM-56 Bulava (SS-N-32)." Missile Threat. https://missilethreat.csis.org/missile/ss-n-32-bulava/

Cybersecurity and Infrastructure Security Agency. (n.d.). "Chinese Malicious Cyber Activity." https://www.cisa.gov/uscert/ncas/current-activity/2020/08/03/chinese-malicious-cyber-activity

Davis, D. L. (2020, July 29). "Responsibly Competing with China." Defense Priorities. https://www.defensepriorities.org/explainers/responsibly-competing-with-china

Davis, M. (2018, July 13). "Space: The Next South China Sea." The Maritime Executive. https://maritime-executive.com/editorials/space-the-next-south-china-sea

Defense Intelligence Agency. (2019, February 11). *Challenges to Security in Space.* Defense Intelligence Agency. https://www.dia.mil/Portals/110/Documents/News/Military_Power_Publications/Challenges_Security_Space_2022.pdf

defensepriorities.org. (2020, June 25). "Burden Shifting to Fix Outdated Alliances." *Defense Priorities.* https://www.defensepriorities.org/explainers/burden-shifting-to-fix-outdated-alliances

Delman, E. (2015, October 2). "The Link between Putin's Military Campaign in Syria and Ukraine." The Atlantic. https://www.theatlantic.com/international/archive/2015/10/navy-base-syria-crimea-putin/408694/

Department of Defense. (2018, February). *2018 Nuclear Posture Review.* https://media.defense.gov/2018/Feb/02/2001872886/-1/-1/1/2018-NUCLEAR-POSTURE-REVIEW-FINAL-REPORT.PDF

Department of Defense. (2020a). *Defense Space Strategy Summary.* Department of Defense. https://media.defense.gov/2020/Jun/17/2002317391/-1/-1/1/2020_DEFENSE_SPACE_STRATEGY_SUMMARY.PDF

Department of Defense. (2020b). *Military and Security Developments Involving the People's Republic of China 2020.* https://media.defense.gov/2020/Sep/01/2002488689/-1/-1/1/2020-DOD-CHINA-MILITARY-POWER-REPORT-FINAL.PDF

Department of Defense. (2022). *Combined Space Operations (CSpO) Vision 2031.* https://media.defense.gov/2022/Feb/22/2002942522/-1/-1/0/CSPO-VISION-2031.PDF

Deplano, R. (2021). *The Artemis Accords: Evolution or Revolution in International Space Law?* Cambridge University Press.

Di Mare, A. (2021, May). *The Role of Space Domain Awareness.* Joint Air & Space Conference 2021. https://www.japcc.org/essays/the-role-of-space-domain-awareness/

Dickey, M. R. (2021, January 31). "New Service, New Architecture: Rising to the Challenge of Delivering Space Force Capabilities." Space Force Journal. https://spaceforcejournal.org/new-service-new-architecture-rising-to-the-challenge-of-delivering-space-force-capabilities/

Dinerman, T. (2021). *Space Force! A Quirky and Opinionated Look at America's Newest Military Service.* Brentwood, TN: Permuted Press.

Dolman, E. C. (2002). *Astropolitik: Classical Geopolitics in the Space Age.* London: Frank Cass & Co.

Donovan, M. (2021, April 27). "It's Imperative America Preserve Its Space Power Advantage." DefenseNews. https://www.defensenews.com/opinion/commentary/2021/04/27/its-imperative-america-preserve-its-space-power-advantage/

Drake, N. (2018, December 20). "Where, Exactly, Is the Edge of Space? It Depends on Who You Ask." National Geographic. https://www.nationalgeographic.com/science/article/where-is-the-edge-of-space-and-what-is-the-karman-line

Dyson, F. (2016). *Freeman Dyson and Project Orion.* American Experience. PBS.

Dyson, G. (2003). *Project Orion: The True Story of the Atomic Spaceship.* New York: Henry Holt and Company.

The Economist. (2019, August 30). "Donald Trump Creates Space Command but Must Wait for Space Force." https://www.economist.com/united-states/2019/08/30/donald-trump-creates-space-command-but-must-wait-for-space-force

Editors of Encyclopedia Britannica. (n.d.). 'Strategic Defense Initiative." Britannica. https://www.britannica.com/topic/Strategic-Defense-Initiative

Ellyatt, H. (2019, December 5). "Putin Fears the US and NATO Are Militarizing Space and Russia Is Right to Worry, Experts Say." CNBC. https://www.cnbc.com/2019/12/05/nato-in-space-putin-is-worried-about-the-militarization-of-space.html

Erickson, A., & Collins, G. (2012, August 30). "China's Real Blue Water Navy." The Diplomat. https://thediplomat.com/2012/08/chinas-not-so-scary-navy/

Erwin, S. (2020a, April 9). "Numerica Expands Space Surveillance Services Aimed at Satellite Operators." SpaceNews. https://spacenews.com/numerica-expands-space-surveillance-services-aimed-at-satellite-operators/

Erwin, S. (2020b, June 4). "Space Force Thinking about NASA-Style Partnerships with Private Companies." SpaceNews. https://spacenews.com/space-force-thinking-about-nasa-style-partnerships-with-private-companies/

Erwin, S. (2021, May 28). "Biden Seeks $2 Billion Funding Boost for U.S. Space Force." SpaceNews. https://spacenews.com/biden-seeks-2-billion-funding-boost-for-u-s-space-force/

European Space Agency. (2022). "Protection of Space Assets." https://vision.esa.int/protection-of-space-assets/

European Space Agency. (n.d.). "Hypervelocity Impacts and Protecting Spacecraft." https://www.esa.int/Space_Safety/Space_Debris/Hypervelocity_impacts_and_protecting_spacecraft

Friedman, G. (2009). *The Next 100 Years.* New York: Anchor Books.

Fukuyama, F. (1992). *The End of History and the Last Man.* New York: Free Press.

Gabuev, A. (2018, February 23). "Russian-U.S. Flashpoints in the Post-Soviet Space: The View from Moscow." Carnegie Moscow Center. https://carnegiemoscow

.org/2018/02/23/russian-u.s.-flashpoints-in-post-soviet-space-view-from-moscow-pub-75631

Gaddis, J. (2005). *Strategies of Containment: A Critical Appraisal of American National Security Policy during the Cold War*. Oxford University Press.

Gallagher, N. (2010, May). "Space Governance and International Cooperation." *Astropoliltics* 8(2). https://www.tandfonline.com/doi/full/10.1080/14777622.2010.524131

Garretson, P. (2021, June). "Demanding More of Space Force." Aerospace America. https://aerospaceamerica.aiaa.org/departments/demanding-more-of-space-force/

Garretson, S. P. (2021, January 19). Interview. (D. L. Colucci, Interviewer).

Gingrich, N. (2019, July 23). "Trump's Plan to Develop the Moon and Mars Will Change Future of Human Race." *Newsweek*. https://www.newsweek.com/trumps-plan-develop-moon-mars-will-change-future-human-race-opinion-1450736

Gingrich, N. (2021, January 30). Interview (D. L. Colucci, Interviewer).

Gleason, M. P. (2021). *Getting the Most Deterrent Value from U.S. Space Forces*. Center for Space Policy Strategy. https://aerospace.org/sites/default/files/2020-10/Gleason-Hays_SpaceDeterrence_20201027_0.pdf

Global Firepower. (2021). "2021 Military Strength Ranking." https://www.globalfirepower.com/countries-listing.php

globalsecurity.org. (n.d.a). "Fractional Orbiting Bombardment Systems (FOBS)." https://www.globalsecurity.org/wmd/world/russia/fobs.htm

globalsecurity.org. (n.d.b). "Russian Paranoia." https://www.globalsecurity.org/military/world/russia/paranoia.htm

globalsecurity.org. (n.d.c) "Strategic Defense Initiative." https://www.globalsecurity.org/space/systems/sdi.htm

Godwin, R. (2019, July 19). "The Forgotten Plans to Reach the Moon—before Apollo." Smithsonian Magazine. https://www.smithsonianmag.com/air-space-magazine/forgotten-plans-reach-moon-apollo-180972695/

Gokhale, N. (2011, January 25). "India's Doctrinal Shift?" The Diplomat. https://thediplomat.com/2011/01/indias-doctrinal-shift/

Goswami, N. (2019, August 5). "China's Grand Strategy in Outer Space: To Establish Compelling Standards of Behavior." The Space Review. https://www.thespacereview.com/article/3773/1

Goswami, N. (2021, January 18). Interview. (D. L. Colucci, Interviewer).

Goswami, N., & Garretson, P. A. (2020). *Scramble for the Skies: The Great Power Competition to Control the Resources of Outer Space*. Lanham, MD: Lexington Books.

Gricius, G. (2020). "Russian Ambitions in the Arctic: What to Expect." Global Security Review. https://globalsecurityreview.com/russia-arctic-ambitions/

Grosselin, K. (2021, January 31). "A Beneficial and Striking Success: Diplomatic Spacepower and Communication Satellites in the Early Space Age (1958–1972)." Space Force Journal. https://spaceforcejournal.org/a-beneficial-and-striking-success-diplomatic-spacepower-and-communication-satellites-in-the-early-space-age-1958-1972/

# Bibliography

Hadley, G. (2021, November 30). "ISR Missions for Space Force 'Just Make a Lot of Sense,' USECAF Says." Air Force Magazine. https://www.airforcemag.com/space-force-isr-missions-makes-sense-usecaf/

Hanlon, M. L., & Autry, G. (2022, January 3). "The Rules of Space Haven't Been Updated in 50 Years, and the UN Says It's Time." CNN. https://www.cnn.com/2022/01/03/world/space-law-united-nations-partner-scn/index.html

Haokip, T. (2014, March 5). "India's Look East Policy: Its Evolution and Approach." *South Asian Survey*. Sage Journals. https://journals.sagepub.com/doi/10.1177/0971523113513368

Harrison, T., Johnson, K., & Roberts, T. G. (2019, April 4). "Space Threat Assessment 2019." Center for Strategic & International Studies. https://www.csis.org/analysis/space-threat-assessment-2019

Hennigan, W. (2020, February 10). "Exclusive: Strange Russian Spacecraft Shadowing U.S. Spy Satellite, General Says." TIME. https://time.com/5779315/russian-spacecraft-spy-satellite-space-force/

Henry, C. (2018, July 20). "Putin Challenges Roscosmos to 'Drastically Improve' on Space and Launch." Spacenews. https://spacenews.com/putin-challenges-roscosmos-to-drastically-improve-on-space-and-launch/

Hentz, J. (2004). *The Obligation of Empire*. Lexington: University of Kentucky Press.

Hill, L. (2019, September 13). "Schriever Wargame Concludes." Air Force Command. https://www.afspc.af.mil/News/Article-Display/Article/1960610/schriever-wargame-concludes/

Hisham, M. (2017, April 10). "How Obama's Syrian Chemical Weapons Deal Fell Apart." The Atlantic. https://www.theatlantic.com/international/archive/2017/04/how-obamas-chemical-weapons-deal-fell-apart/522549/

Hitchens, T. (2016, April 1). "A Pause Button for Militarizing Space." Center for International & Security Studies at Maryland. https://cissm.umd.edu/research-impact/publications/pause-button-militarizing-space

Holmes, J. R., & Yoshihara, T. (2009, December). "Mahan's Lingering Ghost." U.S. Naval Institute. https://www.usni.org/magazines/proceedings/2009/december/mahans-lingering-ghost

Hooker, R. (2014, October). *The Grand Strategy of the United States*. https://ndupress.ndu.edu/Portals/68/Documents/Books/grand-strategy-us.pdf

Howell, E. (2017, August 21). "Lagrange Points: Parking Places in Space." Space.com.https://www.space.com/30302-lagrange-points.html

Howell, E. (2020, February 27). "Trump Hails India's 'Impressive Strides' on Moon Exploration, Pledges Greater Cooperation on Space." Space.com. https://www.space.com/trump-hails-india-moon-missions-us-space-cooperation.html

Hyten, J. E. (2002). "A Sea of Peace or a Theater of War? Dealing with the Inevitable Conflict in Space." Air & Space Power Journal. https://www.airuniversity.af.edu/Portals/10/ASPJ/journals/Chronicles/Hyten.pdf

Impey, C. (2016). *Beyond: Our Future in Space*. New York: W. W. Norton.

Jacobs, E. (2020, August 17). "Russia Offers Statement of Military Support to Belarus' Embattles Leaders." New York Post. https://nypost.com/2020/08/17/russia-offers-military-support-to-belarus-embattled-leader/

Jaeger, P. (2019, November 17). "China's Coup on the Dark Side of the Moon." Financial Times. https://www.ft.com/content/74f63e54-07af-11ea-a984-fbbacad9e7dd

Jaishankar, D. (2019, October 24). "Acting East: India in the Indo-Pacific." Brookings Institution. https://www.brookings.edu/research/acting-east-india-in-the-indo-pacific/

Jayaraman, K. (2015, March 9). "India Allocates $1.2 Billion for Space Activities." Space News. https://spacenews.com/india-allocates-1-2-billion-for-space-activities/

Jefferson, T. (1800). Private letter, quoted in "Thomas Jefferson." Clinton White House Archives. https://clintonwhitehouse4.archives.gov/WH/glimpse/presidents/html/tj3.html

Jenkins, M. (2021, November 1). "Strategic Geographical Points in Outer Space." The Space Review. https://www.thespacereview.com/article/4273/1

Jentleson, B. (2010). *American Foreign Policy*. New York: W.W. Norton.

Jet Propulsion Laboratory. (2022, March 16). "NASA Adds Giant New Dish to Communicate Deep Space Missions." https://www.jpl.nasa.gov/news/nasa-adds-giant-new-dish-to-communicate-with-deep-space-missions

Johnson-Freese, J. (2007). *Space as a Strategic Asset*. New York: Columbia University Press.

Johnson-Freese, J. (2016). *Space Warfare in the 21st Century: Arming the Heavens (Cass Military Studies)*. New York: Taylor & Francis.

Jones, A. (2022, March 6). "China Wants Its New Rocket for Astronaut Launches to Be Reusable." Space.com. https://www.space.com/china-reusable-rockets-for-astronaut-launches

Kaplan, S. (2020, July 13). *Eyes on the Prize: The Strategic Implications of Cislunar Space and the Moon*. Center for Strategic & International Studies. https://aerospace.csis.org/strategic-interest-in-cislunar-space-and-the-moon/

Kapusta, C. P. (2015, September 9). *The Gray Zone*. United States Special Operations Command. https://info.publicintelligence.net/USSOCOM-GrayZones.pdf

Kehler, G. R. (2021, January 28). Interview. (D. L. Colucci, Interviewer).

Kelly, S. (2018, August 23). "Selling Space Force to the American Electorate." The Hill. https://thehill.com/opinion/white-house/403284-selling-space-force-to-the-american-electorate/

Kennedy, J. F. (1962). Address at Rice University on the Nation's Space Effort. John F. Kennedy Presidential Library and Museum. https://www.jfklibrary.org/learn/about-jfk/historic-speeches/address-at-rice-university-on-the-nations-space-effort

Klein, J. (2006). *Space Warfare*. CreateSpace Independent Publishing.

Kliman, D. (2019, April 8). "Grading China's Belt and Road." Center for a New American Security. https://www.cnas.org/publications/reports/beltandroad

Korab-Karpowicz, J. (2010, July 26). "Political Realism in International Relations." Stanford Encyclopedia of Philosophy. https://plato.stanford.edu/entries/realism-intl-relations/

Kramer, A. E., & Myers, S. L. (2021, June 15). "Russia, Once a Space Superpower, Turns to China for Missions." New York Times. https://www.nytimes.com/2021/06/15/world/asia/china-russia-space.html

Kwast, S. (2021, January 15). Interview. (D. L. Colucci, Interviewer).

# Bibliography

Lee, J. (2020, March 4). "Space Force Holds Inaugural ISR-Focused Conference." United States Space Force. https://www.spaceforce.mil/News/Article/2101376/space-force-holds-inaugural-isr-focused-conference/

Lendon, B. (2020, August 21). "Satellite Photos Appear to Show Chinese Submarine Using Underground Base." CNN. https://www.cnn.com/2020/08/21/asia/china-submarine-underground-base-satellite-photo-intl-hnk-scli/index.html

Li, Y. (2019, May 30). "Here's Why China's Trade War Threat to Restrict Rare Earth Minerals Is So Serious." CNBC. https://www.cnbc.com/2019/05/30/heres-why-chinas-trade-war-threat-to-restrict-rare-earth-minerals-is-so-serious.html

Lincoln, A. (1862, December 1). "Annual Message to Congress – Concluding Remarks." Abraham Lincoln Online. https://www.abrahamlincolnonline.org/lincoln/speeches/congress.htm

Lopez, C. T. (2019, March 29). "DOD Official: Maintaining Space Dominance 'Pivotal' for U.S. Warfighters." U.S. Department of Defense. https://www.defense.gov/News/News-Stories/Article/Article/1800891/dod-official-maintaining-space-dominance-pivotal-for-us-warfighters/

Lupton, D. E. (1988, June). *On Space Warfare: A Space Power Doctrine.* Air University Press. https://citeseerx.ist.psu.edu/viewdoc/download?doi=10.1.1.182.6911&rep=rep1&type=pdf

Luttwak, E. (1985). *Strategy and History.* New Brunswick, NJ: Transaction Books.

Mackinder, H. (1904). "The Geographical Pivot of History." The Geographic Journal. https://www.iwp.edu/wp-content/uploads/2019/05/20131016_MackinderTheGeographicalJournal.pdf

macrotrends.net. (n.d.). "U.S. Military Spending/Defense Budget 1960–2021." https://www.macrotrends.net/countries/USA/united-states/military-spending-defense-budget

Mahan, A.T. (1897). *The Interest of America in Sea Power, Present and Future.* London: Sampson Low, Marston, and Company.

Mahan, A. T. (1940). *The Influence of Sea Power upon History.* Boston: Little, Brown, and Company.

Maher, B. W. (2020, April 27). "Preserving Freedom of Action in Cislunar Space." Wild Blue Yonder. https://www.airuniversity.af.edu/Wild-Blue-Yonder/Articles/Article-Display/Article/2159463/preserving-freedom-of-action-in-cislunar-space/

Mainzer, A. (2017, April 20). "The Future of Planetary Defense." AGU. https://agupubs.onlinelibrary.wiley.com/doi/10.1002/2017JE005318

Malachowski, J. (2021, January 31). "Don't Gamble on the Next Space Race: Win in the Orbital Gray Zone Now." Space Force Journal. https://spaceforcejournal.org/dont-gamble-on-the-next-space-race-win-in-the-orbital-gray-zone-now/

Mann, A., Pultarova, T., & Howell, E. (2022, April 14)."SpaceX Starlink Internet: Costs, Collision Risks and How It Works." Space.com. https://www.space.com/spacex-starlink-satellites.html

McCauley, M., & Lieven, D. (n.d.). "Ethnic Relations and Russia's 'Near-Abroad.'" Britannica. https://www.britannica.com/place/Russia/Ethnic-relations-and-Russias-near-abroad

McDougall, W. (2010, April 13). "Can the United States Do Grand Strategy?" Foreign Policy Research Institute. https://www.fpri.org/article/2010/04/can-the-united-states-do-grand-strategy/

McDougall, W. A. (1997). *The Heavens and the Earth: A Political History of the Space Age*. New York: Basic Books.

Mehta, A. (2022, January 13). "Is the Space Force Doing What It's Supposed To?" Breaking Defense. https://breakingdefense.com/2022/01/is-the-space-force-doing-what-its-supposed-to-infographics/

Ministry of Defence of the Russian Federation. (n.d.a). "Space Forces." https://eng.mil.ru/en/structure/forces/cosmic.htm

Ministry of Defence of the Russian Federation. (n.d.b). "Structure: Aerospace Forces." https://eng.mil.ru/en/structure/forces/aerospace.htm

Ministry of Foreign Affairs of the Russian Federation. (2020). "Basic Principles of State Policy of the Russian Federation on Nuclear Deterrence." Hans de Vreij blog. https://hansdevreij.com/2022/03/06/basic-principles-of-state-policy-of-the-russian-federation-on-nuclear-deterrence/

Ministry of Foreign Affairs, the People's Republic of China. (n.d.). "China's Independent Foreign Policy of Peace." https://www.mfa.gov.cn/ce/cegv//eng/zgbd/zgwjzc/t85889.htm

Mishima-Baker, B. A. (2021, March 4). "Moon Wars: Legal Trouble in Space and Moon Law." Digital Reporter Post. https://www.afjag.af.mil/Portals/77/documents/Reporter/20210304%20Mishima%20Baker.pdf?ver=SNGv8IQGp4wlKwQYTAz7Bg%3D%3D

Missile Defense Advocacy Alliance. (2018, August 24). "China's Anti-Access Area Denial." https://missiledefenseadvocacy.org/missile-threat-and-proliferation/todays-missile-threat/china/china-anti-access-area-denial/

Moloney, I. (2021, January 10). Interview. (D. L. Colucci, Interviewer).

Moltz, J. C. (2019, Spring). "The Changing Dynamics of Twenty-First-Century Space Power." Strategic Studies Quarterly. https://www.airuniversity.af.edu/Portals/10/SSQ/documents/Volume-13_Issue-1/Moltz.pdf

Morehouse, Jesse (2022, August 8). "U.S. Space Command Partners with State Partnership Program to Strengthen Space Cooperation." Air National Guard. https://www.nationalguard.mil/News/Article/3119527/us-space-command-partners-with-state-partnership-program-to-strengthen-space-co/

Morgan Stanley. (2020, July 24). "Space: Investing in the Final Frontier." https://www.morganstanley.com/ideas/investing-in-space

Morgenthau, H. (1970). *A New Foreign Policy for the United States*. New York: Praeger.

Morgenthau, H. (1992). *Politics among Nations*. New York: McGraw Hill.

Mozer, J. (2021, March 9). Interview. (D. L. Colucci, Interviewer).

Mueller, K. P. (2010). *Totem and Taboo: Depolarizing the Space Weaponization Debate*. Rand Corporation Reprint. https://www.rand.org/pubs/reprints/RP1076.html

Munoz, C. (2022, January 28). "USAF, Space Force to Draft Joint Requirement for Space-Based Intelligence." Janes. https://www.janes.com/defence-news/news-detail/usaf-space-force-to-draft-joint-requirements-for-space-based-intelligence

Musk, E. (2016, April 29). Tweet. https://twitter.com/elonmusk/status/726216836069515264

NASA. (2018a, January 5). "NASA Deep Space Exploration Systems Look Ahead to Action-Packed 2018." https://www.nasa.gov/feature/nasa-deep-space-exploration-systems-look-ahead-to-action-packed-2018

NASA. (2018b, May 2). "NASA's Lunar Outpost Will Extend Human Presence in Deep Space." https://www.nasa.gov/feature/nasa-s-lunar-outpost-will-extend-human-presence-in-deep-space

NASA. (2019a). "NASA's Plan for Sustained Lunar Exploration and Development." https://www.nasa.gov/sites/default/files/atoms/files/a_sustained_lunar_presence_nspc_report4220final.pdf

NASA. (2019b). "Solar System Exploration." https://solarsystem.nasa.gov/

NASA. (2020). *Memorandum of Understanding between the National Aeronautics and Space Administration and the United States Space Force.* https://www.nasa.gov/sites/default/files/atoms/files/nasa_ussf_mou_21_sep_20.pdf

NASA. (2021, January 27). "NASA's Artemis Base Camp on the Moon Will Need Light, Water, Elevation." https://www.nasa.gov/feature/goddard/2021/nasa-s-artemis-base-camp-on-the-moon-will-need-light-water-elevation

NASA. (n.d.). "Space Shuttle Era." https://www.nasa.gov/mission_pages/shuttle/flyout/index.html

NASA/WMAP Science Team. (2018, March 27). "What Is a Lagrange Point?" NASA. https://solarsystem.nasa.gov/resources/754/what-is-a-lagrange-point/

National Commission on Terrorist Attacks upon the United States. (2004). *Final Report of the National Commission on Terrorist Attacks upon the United States.* https://www.govinfo.gov/content/pkg/GPO-911REPORT/pdf/GPO-911REPORT.pdf

National WW2 Museum. (n.d.). "Research Starters: Worldwide Deaths in World War II." https://www.nationalww2museum.org/students-teachers/student-resources/research-starters/research-starters-worldwide-deaths-world-war

NESDIS. (2022, March 14). "NOAA's GOES-T Reaches Geostationary Orbit, Now Designated GOES-18." https://www.nesdis.noaa.gov/news/noaas-goes-t-reaches-geostationary-orbit-now-designated-goes-18

Newman, C. (2020, October 19). "Artemis Accords: Why Many Countries Are Refusing to Sign Moon Exploration Agreement." The Conversation. https://theconversation.com/artemis-accords-why-many-countries-are-refusing-to-sign-moon-exploration-agreement-148134

Noll, J., Bojang, O., & Sebastiaan, R. (2020, January). "Deterrence by Punishment or Denial? The eFP Case." In F. Osinga and T. Sweijs, eds., *NL ARMS Netherlands Annual Review of Military Studies 2020*, pp. 109–128. The Hague: NL ARMS, T.M.C. Asser Press. https://link.springer.com/chapter/10.1007/978-94-6265-419-8_7

The Obama White House. (2015, September 28). *Remarks by President Obama to the United Nations General Assembly.* https://obamawhitehouse.archives.gov/the-press-office/2015/09/28/remarks-president-obama-united-nations-general-assembly

Oberg, J. E. (1999). *Space Power Theory.* Colorado Springs, CO: U.S. Air Force Academy.

Odell, L. A., DiLorenzo, C. D., Dawson, C. A., & DeMaio, W. A. (2021, June). *U.S. Low Earth Orbit Dominance Shifting with Gray Zone Competition.*

Institute for Defense Analyses. https://www.ida.org/-/media/feature/publications/u/us/us-low-earth-orbit-dominance-shifting-with-gray-zone-competition/d-22676.ashx

Office of the Director of National Intelligence. (2021, April 13). *Annual Threat Assessment of the U.S. Intelligence Community*. https://www.dni.gov/index.php/newsroom/reports-publications/reports-publications-2021/item/2204-2021-annual-threat-assessment-of-the-u-s-intelligence-community

Office of the Press Secretary. (2019). *Space Policy Directive-4: Establishment of the United States Space Force*. United States Space Force. https://www.spaceforce.mil/About-Us/SPD-4/

O'Hanlon, M. E. (2004). *Neither Star Wars nor Sanctuary: Constraining the Military Uses of Space*. Washington, DC: Brookings Institution Press.

Ohlandt, C. J., McClintock, B., & Flanagan, S. J. (2021, February 7). "Navigating Norms for the New Space Era." The National Interest. https://nationalinterest.org/feature/navigating-norms-new-space-era-177592

O'Neill, A. (2022). "Number of United States Military Fatalities in Major Wars 1775–2022." Statista. https://www.statista.com/statistics/1009819/total-us-military-fatalities-in-american-wars-1775-present/

O'Neill, G. K. (2019). *The High Frontier: Human Colonies in Space: Apogee Books Space Series 12*. Burlington, Ontario: Collector's Guide Publishing.

Pant, H. V. (2019, June 4). "Modi Reimagines India's Role in the World." Foreign Policy. https://foreignpolicy.com/2019/06/04/modi-reimagines-indiasrole-in-the-world/

Peter, T. (2019, April 17). "Back on Earth, China's Mars simulation base Greets First Visitors." Reuters. https://www.reuters.com/article/us-space-exploration-china-mars/back-on-earth-chinas-mars-simulation-base-greets-first-visitors-idUSKCN1RT11S

Peterson Space Force Base. (n.d.). *Space Base Delta 1*. https://www.spacebasedelta1.spaceforce.mil/Peterson-SFB-Colorado/

Piesing, M. (2021, January 22). "Spaceplanes: The Return of the Reusable Spacecraft?" BBC. https://www.bbc.com/future/article/20210121-spaceplanes-the-return-of-the-reuseable-spacecraft

Pollpeter, K. (2021). *Space Domain Awareness as a Strategic Counterweight*. China Strategic Studies Institute. https://www.cna.org/archive/CNA_Files/pdf/space-domain-awareness-as-a-strategic-counterweight.pdf

Poole, R. E. (2014). "China's 'Harmonious World' in the Era of the Rising East." *Inquiries Journal* 6(10). http://www.inquiriesjournal.com/a?id=932

President of Russia. (2007, February 10). "Speech and the Following Discussion at the Munich Conference on Security Policy." http://en.kremlin.ru/events/president/transcripts/24034

Prisco, G. (2020, September 14). "The West Needs Bold, Sustainable, and Inclusive Space Programs and Visions, or Else." The Space Review. https://www.thespacereview.com/article/4023/1

Pultarova, T. (2021, June 17). "Russia, China Reveal Moon Base Roadmap but No Plans for Astronaut Trips Yet." Space.com. https://www.space.com/china-russia-international-lunar-research-station

Raju, N. (2021, August 27). "How Is Lunar Security Different from Space Security?" Moon Dialogs. https://www.moondialogs.org/2021-approach/lunar-security-different-space-security?rq=Raju%20Lunar%20Security

# Bibliography

Raymond, G. J. (2021, June 16). Interview. (D. L. Colucci, Interviewer).

research outreach. (2020, November 4). "Planetary Protection Policy: For Sustainable Space Exploration and to Safeguard Our Biosphere." https://researchoutreach.org/articles/planetary-protection-policy-sustainable-space-exploration-safeguard-biosphere/

Reynolds, M. (2019, October 24). "Turkey and Russia: A Remarkable Rapprochement." War on the Rocks. https://warontherocks.com/2019/10/turkey-and-russia-a-remarkable-rapprochement/

Rhaguvanshi, V. (2019, June 12). "India to Launch a Defense-Based Space Research Agency." Defense News. https://www.defensenews.com/space/2019/06/12/india-to-launch-a-defense-based-space-research-agency/

Rieves, A. (2020, August 21). "First Space Force Intelligence Officer to Train at GAFB." Goodfellow Air Force Base. https://www.goodfellow.af.mil/Newsroom/Article-Display/Article/2321196/first-space-force-intelligence-officer-to-train-at-gafb/

Roblin, S. (2020, May 13). "China Touts New Submarine—Launched Ukes in Quest for More Survivable Deterrence." Forbes. https://www.forbes.com/sites/sebastienroblin/2020/05/13/china-touts-new-submarine-launched-nukes-in-quest-for-more-survivable-deterrence/?sh=28b18f917554

Rogers, M. (2017). *Remarks to 2017 Space Symposium*. U.S. Government.https://www.defensedaily.com/wp-content/uploads/post_attachment/163165.pdf

Roscosmos. (2018, November 28). *Joint Meeting of the Scientific and Technical Council of Roscosmos and the Space Council of the Russian Academy of Science.* https://www.roscosmos.ru/25789/.%c2%a0/

Rosenbaum, E., & Donovan, R. (2019, March 17). "China Plans a Solar Power Play in Space That NASA Abandoned Decades Ago." CNBC. https://www.cnbc.com/2019/03/15/china-plans-a-solar-power-play-in-space-that-nasa-abandoned-long-ago.html

Rossettini, L. (2021, April 13). "Space Infrastructure Is the Next Investment Frontier and SPACs Are a Launch Pad." Marketwatch. https://www.marketwatch.com/story/space-infrastructure-is-the-next-investment-frontier-and-spacs-are-a-launch-pad-11618294752

Rothman, N. (2014, March 3). "Flashback: In Russia, Obama Declared 'Great Power Conflict' a Thing of the Past." Mediaite. https://www.mediaite.com/online/flashback-in-russia-obama-declared-great-power-conflict-a-thing-of-the-past/

Royal Museums Greenwich. (n.d.). "Inside NASA's Artemis Mission." https://www.rmg.co.uk/stories/topics/nasa-moon-mission-artemis-program-launch-date

Royde-Smith, J. (n.d.). "World War I: Killed, Wounded, and Missing." Britannica. https://www.britannica.com/event/World-War-I/Killed-wounded-and-missing

RT. (2014, May 13). "Russia's Black Sea Fleet to Receive 30 New Ships, Become Self-Sufficient." https://www.rt.com/news/158772-black-sea-fleet-ships/

Sacks, D. (2019, November 1). "China's Efforts to Diplomatically Isolate Taiwan Could Backfire." The National Interest. https://nationalinterest.org/feature/chinas-efforts-diplomatically-isolate-taiwan-could-backfire-92886

SARSAT. (2020). *Interagency Memorandum of Agreement for the United States Satellite Aided Searchand Rescue System.* NASA. https://www.nasa.gov/saa/domestic/35144_SARSAT_MOA_2020_Fully_Signed_All_Parties(1).pdf
Satherly, D. (2020, September 16). "China Military Moving Away from 'No-First-Use' Nuke Policy—US Commander's Stark Warning." Newshub. https://www.newshub.co.nz/home/world/2020/09/china-military-moving-away-from-no-first-use-nuke-policy-us-commander-s-stark-warning.html
Schaffer, A. M. (2017, October 1). *The Role of Space Norms in Protection and Defense.* https://ndupress.ndu.edu/Portals/68/Documents/jfq/jfq-87/jfq-87_88-92_Schaffer.pdf?ver=2017-09-28-092555-747
Seck, H. (2022, March 21). "Why Space Force Wants to Patrol around the Moon." Sandboxx. https://www.sandboxx.us/blog/why-space-force-wants-to-patrol-around-the-moon/
Secretary of the Air Force Public Affairs. (2020, May 6). "Next X-37B Orbital Test Vehicle Scheduled to Launch." United States Space Force. https://www.spaceforce.mil/News/Article/2177702/next-x-37b-orbital-test-vehicle-scheduled-to-launch/
Secretary of the Air Force Public Affairs. (2021, August 23). "Space Force Activates Space Training and Readiness Command." United States Space Force. https://www.spaceforce.mil/News/Article/2742956/space-force-activates-space-training-and-readiness-command/
Seminari, S. (2019, November 24). "Global Government Space Budgets Continues Multiyear Rebound." SpaceNews. https://spacenews.com/op-ed-global-government-space-budgets-continues-multiyear-rebound/
Sharma, Ashok. (2022). "India Launches New Aircraft Carrier as China Concerns Grow." DefenseNews. https://www.defensenews.com/naval/2022/09/02/india-launches-new-aircraft-carrier-as-china-concerns-grow/
Shatner, B. (2020, August 26). "William Shatner Wants to Know: What the *Heck Is Wrong* with You, Space Force?" Military Times. https://www.militarytimes.com/opinion/commentary/2020/08/26/what-the-heck-is-wrong-with-you-space-force/
Shaw, J. (2021, February 23). Interview. (D. L. Colucci, Interviewer).
Singh, S. (2020, September 14). "Can India Transcend Its Two-Front Challenge?" War on the Rocks. https://warontherocks.com/2020/09/can-india-transcend-its-two-front-challenge/
Smith, C. (2017, March 13). "America Needs a Space Corps." The Space Review. https://www.thespacereview.com/article/3193/1
Smith, C. (2020, December 20). Interview. (D. L. Colucci, Interviewer).
Smith, M. (2020, May 5). "NASA and Space Force to Work Together on Planetary Defense."spacepolicyonline.com. https://spacepolicyonline.com/news/nasa-and-space-force-to-work-together-on-planetary-defense/
The Space Report. (2021). "Space Force: Mission Moves Forward as New Branch Begins Second Year." https://www.thespacereport.org/uncategorized/space-force-mission-moves-forward-as-new-branch-begins-second-year/
SpaceX. (2022). "Falcon 9." https://www.spacex.com/vehicles/falcon-9/

# Bibliography

Stanzel, A. (2019, June 25). "China Trends #2—China's String of Ports in the Indian Ocean." Institut Montaigne. https://www.institutmontaigne.org/en/blog/china-trends-2-chinas-string-ports-indian-ocean

STARCOM. (n.d.). "Space Training and Readiness Command." https://www.starcom.spaceforce.mil/

Starling, C. G., Massa, M. J., Mulder, C. P., & Siegel, J. T. (2021, April 11). "The Future of Security in Space: A Thirty-Year US Strategy." Atlantic Council Strategy Papers. https://www.atlanticcouncil.org/content-series/atlantic-council-strategy-paper-series/the-future-of-security-in-space/

Stone, M. (2021, December 13). "U.S. Space Force Holds War Game to Test Satellite Network Under Attack." Reuters. https://www.reuters.com/business/aerospace-defense/us-space-force-holds-war-game-test-satellite-network-under-attack-2021-12-14/

Stratfor. (2014, February 19). "A Chronology of Russia's Military Defense Strategy." https://worldview.stratfor.com/article/chronology-russias-military-and-defense-strategy

Suri, M., & Gupta, S. (2019, September 5). "India's Polar Moon Mission Puts Chandrayaan-2 in the History Books." CNN. https://www.cnn.com/2019/09/04/world/india-moon-lunar-chandrayaan-2-explainer-scn/index.html

Swisher. (2021, March 8). "I Asked the Head of Space Force What the Agency Has Done for Me Lately." New York Times. https://www.nytimes.com/2021/03/08/opinion/sway-kara-swisher-john-raymond.html

Szymanski, P. (2020, June 22). "Issues with the Integration of Space and Terrestrial Military Operations." Wild Blue Yonder Journal. https://www.airuniversity.af.edu/Wild-Blue-Yonder/Article-Display/Article/2226268/issues-with-the-integration-of-space-and-terrestrial-military-operations/

Tadjdeh, Y. (2022, January 18). "Space Force Submitting 'Bold' Budget Request for 2023." National Defense Magazine. https://www.nationaldefensemagazine.org/articles/2022/1/18/space-force-submitting-bold-budget-request

Thompson, A. (2017, November 16). "China Wants a Nuclear Space Shuttle by 2040." Popular Mechanics. https://www.popularmechanics.com/space/rockets/a13788331/chinas-future-space-plans/

Tkacik, J. J., Jr. (2005, February 21). "China Is Using North Korea as Leverage." Wall Street Journal. https://www.wsj.com/articles/SB110895247121759897

Tonar, R., & Talton, E. (2018, December 10). "To Commercialize Space We Need to Build Infrastructure, Not Just Launch Rockets." Forbes. https://www.forbes.com/sites/ellistalton/2018/12/10/to-commercialize-space-we-need-to-build-infrastructure-not-just-launch-rockets/?sh=7cb375f01318

Toucas, B. (2017, June 28). "Russia's Design in the Black Sea: Extending the Buffer Zone." Center for Strategic & International Studies. https://www.csis.org/analysis/russias-design-black-sea-extending-buffer-zone

Trevithick, J. (2020, June 9). "Air Force Is Looking beyond Traditional Orbits to Get an Upper Hand in Space." The War Zone Wire. https://www.thedrive.com/the-war-zone/33888/air-force-is-looking-beyond-traditional-orbits-to-get-an-upper-hand-in-space

Tumlinson, R. (2020, December 16). Interview. (D. L. Colucci, Interviewer).

U.S. Air Force. (n.d.). "X-37B Orbital Test Vehicle." https://www.af.mil/About-Us/Fact-Sheets/Display/Article/104539/x-37b-orbital-test-vehicle/

U.S.-China Economic and Security Review Commission. (2019, April 25). *Hearing on China in Space: A Strategic Competition?* https://www.uscc.gov/hearings/china-space-strategic-competition

U.S. Department of Defense. (2021, January 8). "USSF Becomes 18th Member of Intel Community." https://www.defense.gov/News/Releases/Release/Article/2466657/ussf-becomes-18th-member-of-intel-community/

U.S. Department of Defense. (2022, February 22). "DOD and Partners Release Combined Space Operations Vision 2031." https://www.defense.gov/News/Releases/Release/Article/2941594/dod-and-partners-release-combined-space-operations-vision-2031/

U.S. Department of State. (2019). *2019 Country Reports on Human Rights Practices: China (Includes Hong Kong, Macau, and Tibet)—Hong Kong.* https://www.state.gov/reports/2019-country-reports-on-human-rights-practices/china/

U.S. Government Accountability Office. (2021, June 3). "To the Moon! What Challenges Does NASA's 2024 Lunar Mission Face?" https://www.gao.gov/blog/moon-what-challenges-does-nasas-2024-lunar-mission-face

U.S. Space Command. (2019, September 5). *The Future of Space 2060 and Implications for U.S. Strategy: Report on the Space Futures Workshop.* https://aerospace.csis.org/wp-content/uploads/2019/09/Future-of-Space-2060-v2-5-Sep.pdf

U.S. Space Force. (2020). *Spacepower: Doctrine for Space Forces.* https://www.spaceforce.mil/Portals/1/Space%20Capstone%20Publication_10%20Aug%202020.pdf

U.S. Space Force. (2021, December 6). "2021 USSF Campaign Support Plan." https://www.spaceforce.mil/News/Article/2862736/2021-ussf-campaign-support-plan/

U.S. Space Force. (2022). "USSF Mission." https://www.spaceforce.mil/About-Us/About-Space-Force/Mission/

Underwood, K. (2021, April 27). "Space Force Prepares to Launch National Space Intelligence Center." Signal. https://www.afcea.org/signal-media/space-force-prepares-launch-national-space-intelligence-center

United Nations. (1967). *Resolution Adopted by the General Assembly.* https://www.unoosa.org/oosa/en/ourwork/spacelaw/treaties/outerspacetreaty.html

United Nations. (2020). *The Artemis Accords.* https://www.nasa.gov/specials/artemis-accords/img/Artemis-Accords-signed-13Oct2020.pdf

United Nations. (2021, November 1). "Delegates Approve 5 Draft Resolutions, as First Committee Takes Action on Peaceful Use, Non-Weaponization of Outer Space, Chemical Weapons." United Nations. https://press.un.org/en/2021/gadis3676.doc.htm

United States. (2006). *U.S. National Space Policy.* https://activityinsight.pace.edu/tbarnet/intellcont/UNSNSP%2020062012-1.pdf

United States. (2011, January 3). *U.S. National Security Space Strategy.* https://www.dni.gov/index.php/newsroom/reports-publications/reports-publications-2011/item/620-national-security-space-strategy

United States Army. (1959). *Project Horizon.* https://nsarchive2.gwu.edu/NSAEBB/NSAEBB479/docs/EBB-Moon01_sm.pdf

Urrutia, D. E. (2019, March 30). "India's Anti-Satellite Missile Test Is a Big Deal. Here's Why." Space.com. https://www.space.com/india-anti-satellite-test-significance.html

Vedda, J. A. (2018, April). *Cislunar Development: What to Build—and Why*. Center for Space Policy and Strategy. https://aerospace.org/sites/default/files/2018-05/CislunarDevelopment.pdf

Venable, J. (2021). *Rebuilding America's Military: The United States Space Force*. Washington, DC: The Heritage Foundation.

Wall, M. (2019, January 5). "China Just Landed on the Moon's Far Side—and Will Probably Send Astronauts on Lunar Trips." Space.com. https://www.space.com/42914-china-far-side-moon-landing-crewed-lunar-plans.html

Wallace, D. (2015, December 4). "The Church under Putin: Nationalism and Russian Orthodoxy." The Christian Century. https://www.christiancentury.org/article/2015-11/church-under-putin

The White House. (1978, May 11). *Presidential Directive/NSC-37*, "National Space Policy." https://spp.fas.org/military/docops/national/nsc-37.htm

The White House. (1982, July 4). *National Security Decision Directive Number 42*, "National Space Policy." https://www.hq.nasa.gov/office/pao/History/nsdd-42.html

The White House. (1988, February 11). *Presidential Directive on National Space Policy*. https://spp.fas.org/military/docops/national/policy88.htm

The White House. (2017, December 11). *Presidential Memorandum on Reinvigorating America's Human Space Exploration Program*. https://trumpwhitehouse.archives.govpresidential-actions/presidential-memorandum-reinvigorating-americas-human-space-exploration-program/

The White House. (2018, March 23). *President Donald J. Trump Is Unveiling an America First National Space Strategy*. https://aerospace.csis.org/wp-content/uploads/2018/09/Trump-National-Space-Strategy.pdf

The White House. (2021, December). *United States Space Priorities Framework*. https://www.whitehouse.gov/wp-content/uploads/2021/12/United-States-Space-Priorities-Framework-_-December-1-2021.pdf

Whiting, T. (2020, July 23). "USSF Graduates First Candidates of Space Intelligence Intern Program." Space Force News. https://www.spaceforce.mil/News/Article/2286124/ussf-graduates-first-candidates-of-space-intelligence-intern-program/

Wirtz, J. J. (2009). *Space and Grand Strategy*. New York: Routledge.

Wood, T. (2020, October 23). "Who Owns Our Orbit: Just How Many Satellites Are There in Space?" World Economic Forum. https://www.weforum.org/agenda/2020/10/visualizing-easrth-satellites-sapce-spacex/

Woolf, A. F. (2020). *Russia's Nuclear Weapons: Doctrine, Forces, and Modernization*. Congressional Research Service. https://crsreports.congress.gov/product/details?prodcode=R45861

Woolf, A. F. (2021). *Nonstrategic Nuclear Weapons*. Congressional Research Service. https://crsreports.congress.gov/product/details?prodcode=RL32572

Woolsey, J. (2021, February 8). Interview. (D. L. Colucci, Interviewer).

Woomera. (2018). *The Woomera Manual on the International Law of Military Space Operations*. https://law.adelaide.edu.au/woomera/system/files/docs/Woomera%20Manual.pdf

Worden, G. P. (2021, January 18). Interview. (D. L. Colucci, Interviewer).
YouGov. (2018, August 13). "Is the Space Force a Good Idea? What's Better for Students—A Four-Day Week or a Five-Day Week? Have You Eaten Insects?" https://today.yougov.com/opi/surveys/results/#/survey/25cedc9f-9ca9-11e8-824b-1167078a2492
Ziarnick, B. D. (2015). *Developing National Power in Space*. Jefferson, NC: McFarland & Company.
Ziarnick, B. D. (2021, January 31). "A Practical Guide for Spacepower Strategy." Space Force Journal. https://spaceforcejournal.org/a-practical-guide-for-spacepower-strategy/
Zivitski, L. (2020, June 23). "China Wants to Dominate Space, and the US Must Take Countermeasures." DefenseNews. https://www.defensenews.com/opinion/commentary/2020/06/23/china-wants-to-dominate-space-and-the-us-must-take-countermeasures/

# Index

American power struggle, 1–2
   vital interests, 1
   Western diplomatic efforts, 1
America's reactive nature, 45–47
Apollo program, 37
Arnold, David, 54–55
Artemis Accords, 61–62, 112, 123
Artemis program, 125–127
Astropolitics, 14–15, 63–64, 144–145, 153–155, 160–163
Astrostrategy, 51, 110

Bowen, Bleddyn, 21

Carlson, Josh, 19–20, 38, 48, 64, 70–71
Cislunar space, 15, 67, 107–108
Cold war and space, 32–33
Competition with other great powers, 76–78
   great powers and conflict, 76–77
   greatest threats in space, 78
Cooper, Hank, 5, 33–34

Democracy and democratic values, 24
   liberty promotion, 100–102

Department of Defense, 4, 23, 49, 54, 67, 78
Dulles, Allen, 53

Economic strategy and thinking, 67–68
   resources, 68–69

Fukuyama, Francis, 77
Future space conflict, 79–80
Futures cone, 6

Garretson, Simon Pete, 4, 19, 42, 69, 71, 75–76, 88, 90
Geocentrism, 21, 91–92
Geopolitics, 14–15, 18–21, 41, 44–45, 64–65
Geostrategy, 18–19, 103, 110
Gingrich, Newt, 4, 37, 63, 70, 73, 95
Global norms in space, 59–61
   international law, 60–61
   positivist norm, 62–63
Goswami, Namrata, 19, 42, 69–71, 75, 86–87, 90

Grand strategy, 1–2, 14–15, 18, 41–52
   American grand strategy, 45–47, 51
   grand Strategic thinking, 41, 44–45
   national interests, 2–3
   peripheral interests, 3

Historical space intelligence, 53
Horizon project case study, 35–36

ICBM case study, 35
India space doctrine, 88–90
Intelligence community, 53–54
Intelligence for space, 54
Interests, 2–4

Kehler, Robert, 5, 47, 93
Kwast, Steven, 5, 55–56, 63–64, 72, 128, 140–143

Ladder—effective national security, 42–44
Lagrange points, 20, 67, 108–109, 126

Mahan, Alfred Thayer, 19–20, 26–27, 74, 76
Militarization of space, 31–32
Mozer, Joel, 47–48, 63, 72, 78, 93
Mueller, Karl, 31–32

National Security Act of 1947, 23–24
National security doctrines of the American Presidency, 41–42
   black water, 19, 38, 96, 105
   blue water, 19–20, 20–21, 41–42, 47–48
   brown water, 19, 42
   green water, 19, 55
   naval waters, 19–21, 38–41, 47–48, 96–97
Navy of space, 18–20
New space, 15
New space companies, 69–70
Nongovernmental enterprises, 78–79
NSDD 42, 37

O'Neill, Gerard, 21, 69, 126
Orbits, 15
   GEO, 15
   Karman line, 15
   LEO, 15
Orion Project case study, 36
Outer Space Treaty, 35, 61–62, 123–124

Pax Americana, 59–60, 104
Pax Astra, 2
   national space strategy, 2
Political shortsightedness, 25–26
Power cycle, 46
Previous hegemon's grand strategy, 44
Putin's Russian space doctrine, 79–83
   Soviet grand strategy, 82–83

Raymond, Gen. John, 16, 47, 63, 83, 91–92
Required future intelligence, 55–56
   Molony, Ian, 56–57
   U.S. Space Intelligence Service, 55–56, 127
Rogers, Mike, 16

Schriever, Bernard, 76
Security in space, 70–72
Shaw, John, 5, 47–48, 57, 64, 71, 93–94
Situational awareness (LEO), 55–57
Smith, Coyote, 3–4, 5, 19, 38–39, 47–48, 55, 42, 83–84, 93–95
Space alliances, 63–65
   Democratic Alliance of Nations, 64–65
Space control, 5–13, 17, 20–22, 44–46
Space doctrine: Second half of 21st century (long-term), 105–117
   beyond cislunar space, 107–108
   human protection, 115–116, 129
   law and order, 112
   prevention of foreign threats, 110
   space domain awareness (SDA), 112–113
   space dominance, 105–106
   space governance, 106–107, 113–115
   USSF fleet, 110–111

Index

Space doctrine: Upcoming decades (mid-term), 117–130
   defense system in space, 128
   LEO, GEO, and xGEO, 124–125
   lunar presence, 125–126
   protection of space assets, 118–119, 126–127
   USSF and private sector, 120–121, 126
   USSF and the U.S. electorate, 117–118
   war games and simulations, 121–122
Space factions, 20–22
   military faction, 20
   scientific faction, 20
   vision of space, 21–22
Space Force doctrine, 99–104, 122–123
Space Force mission, 94–96
Space Futures workshop, 5–13
Space impact on humanity, 50–51
   American efforts, 51
Space in 21st century, 37–39
   Bush administration, 37–38
   Obama administration, 38
   Trump administration, 39
Space law enforcement, 56–57
Space militarization, 22–23
Space power, 45–46, 50
   Lupton, Colonel David, 45
   Oberg, James, 45
Space strategy, 47–51
   Phase 1, 47–48
   Phase 2, 48–49

Space and national security, 4–5, 120
   offset strategy, 5
Space and science fiction, 27–31
   other fiction and non-fiction works, 31
   *Star Trek*, 27–28
   *Star Wars*, 28–30
Space weapons, 33–36
   anti-satellite weapons, 33–34
   Strategic Defense Initiative (SDI), 35–36
   weapons of mass destruction, 34
Strategy 2040, 96–97

Triplanetary project, 72–74
Trudeau, Arthur, 35–36
Tumlinson, Rick, 15, 69

U.S. space problems, 16–17
USSF independence, 91–94
USSF motto, 105–106

Woolsey, Jim, 53, 82, 94
Worden, Pete, 15, 19, 48

Xi Jinping and China's space doctrine, 84–88
   Belt and Road Initiative (BRI), 85–86
   Chinese Space Force, 86–88

Ziarnick, Brent, 16, 22, 33–34, 36, 45–46, 55, 70, 96

**About the Author**

**Lamont C. Colucci**, PhD, is the inaugural director of Doctrine Development for the U.S. Space Force. He has experience as a diplomat with the U.S. Department of State and is today a full professor of political science at Concordia University. His primary area of expertise is U.S. national security and U.S. foreign policy. He teaches national security, foreign policy, intelligence, terrorism, and international relations courses. He has published three books as the sole author: *Crusading Realism: The Bush Doctrine and American Core Values after 9/11*; *The National Security Doctrines of the American Presidency: How They Shape Our Present and Future* (two volumes), and *The International Relations of the Bible*. He is contributing author to *The Impact of 9/11 on Politics and War: The Day That Changed Everything?*, and *Homeland Security and Intelligence*.

In 2012, Dr. Colucci became the Fulbright Scholar in Residence at the Diplomatic Academy in Vienna, Austria. He has undergraduate and graduate degrees from the University of Wisconsin-Madison and a doctorate in politics from the University of London, England. In 2007, he received Ripon College's Severy Excellence in Teaching Award and, in 2010, the Underkofler Outstanding Teaching Award. In 2015, he received the National Significant Sig award of the Sigma Chi Fraternity. Dr. Colucci is a substantial contributor to U.S. Air Force/Space Command/Space Force publications and conferences concerning space strategy and is an occasional columnist for the *Washington Times*, *National Review*, *Washington Examiner/Weekly Standard*, *The Hill*, *U.S. News and World Report*, and *Defense News*. He is a bimonthly columnist for *Newsmax*.

Dr. Colucci is a Senior Fellow in National Security Affairs at the American Foreign Policy Council and Senior Advisor in National Security for Contingent Security. He is a National Security and Foreign Affairs Advisor to the NATO-oriented Conference of Defence Associations Institute. Dr. Colucci also teaches in the graduate program in intelligence and security at American Military University. He is cofounder and CEO of Space Fund Intelligence and served as the founding interim director of the Center for Politics at Ripon College. In 2018, he was appointed to the National Task Force on National and Homeland Security.